Liquid-Phase Epitaxial Growth of III–V Compound Semiconductor Materials and their Device Applications

Liquid-Phase Epitaxial Growth of III–V Compound Semiconductor Materials and their Device Applications

M G Astles

Royal Signals and Radar Establishment, Malvern

Adam Hilger
Bristol, Philadelphia and New York

。4219296

PHYSICS

© IOP Publishing Ltd 1990

British Library Cataloguing in Publication Data
Astles, M. G.
 Liquid-phase epitaxial growth of III-V compound
 semiconductor materials and their device applications.
 1. Semiconductors
 537.622
 ISBN 0-7503-0044-2

Library of Congress Cataloging-in-Publication Data
Astles, M. G.
 Liquid-phase epitaxial growth of III-V compound
semiconductor materials and their device applications/M. G.
Astles
  p.   cm.
 Includes bibliographical references and index.
 ISBN 0-7503-0044-2
 1. Compound semiconductors. 2. Epitaxy.
 3. Crystals–Growth. 4. Phase rule and equilibrium. I. Title.
 QC611.8.C64A84 1990
 537.6'22–dc20   90-37767
                                                        CIP

Published under the Adam Hilger imprint by IOP Publishing Ltd
Techno House, Redcliffe Way, Bristol BS1 6NX, England
335 East 45th Street, New York, NY 10017-3483, USA
US Editorial Office: 1411 Walnut Street, Philadelphia, PA 19102

Typeset by KEYTEC, Bridport, Dorset
Printed in Great Britain by Galliard (Printers) Ltd, Norfolk

Contents

v

Preface

In spite of the large volume of research on the application of liquid-phase epitaxy (LPE) to the growth of the III–V family of materials, there has been no book covering all the compounds and alloys grown by LPE. There have been book chapters and numerous review articles covering certain aspects of LPE growth or specific III–V compounds (see the Bibliography), but I believe that there has been no attempt to cover the broad sweep of III–V materials and devices. Obviously, in such a broad treatment it is not possible to go into great detail on all subjects. As far as the materials are concerned, most interest, and hence research, has been on GaP, GaAs, (GaAl)As, InP, (GaIn)As and (GaIn)(AsP). Thus most examples of properties and device applications will concentrate on these compounds and alloys. On the subject of device applications, LPE material has largely been used for opto-electronic devices such as lasers, light emitting diodes, photocathodes and solar cells, and so the emphasis in this book is on these devices.

The technique of LPE has very much grown up with the increase in interest in the III–V materials from the early single-layer or simple homojunction devices to the interest in lattice-matched heterojunctions with GaAs/(GaAl)As or InP/(GaIn)As, and finally to the lattice-matched quaternary alloys such as (GaIn)(AsP) on InP or GaAs, (GaAl)(AsSb) on GaSb and (GaIn)(AsSb) on GaSb or InAs. In parallel with this, the demands for higher performance and more reliable devices have put continual pressure on crystal growers to improve their understanding and control of the growth process. The fact that LPE is an 'equilibrium' process has meant that such understanding and control has been possible by the application of classical equilibrium thermodynamics and diffusion theory. Ironically, it has been the equilibrium nature of LPE which has led to its partial demise during the 1980s since many novel device concepts such as quantum wells, strained-layer superlattices and GaAs-on-Si have required non-equilibrium growth processes such as MOVPE or MBE. I believe in the old saying about 'horses for courses'; in other words that there is a role for all growth techniques depending on the type of device, its required performance and cost.

I hope that this book will be of use not only to the existing practitioners of LPE, but to those with an interest in the device applications of the III–V materials and to those studying materials science or applied physics at undergraduate or postgraduate level.

Finally, I am often reminded of a phrase used by a colleague of mine when I first began work on the III–V's in the early 1970s: "If it can't be done in silicon, then it's not worth doing!" I doubt if that was true then—it is certainly not true now in 1990, as a variety of commercially important III–V devices have become established, even if they represent only a small fraction of the silicon-dominated market in electronic devices. The wide range of properties available in the III–V family of materials ensures them a role in many future device applications, and I am sure that liquid-phase epitaxy will continue to have an important role to play for many years to come.

M G Astles
RSRE Malvern
March 1990

Acknowledgments

Firstly, my thanks go to the many friends and colleagues with whom I have worked at Oxford University, the Services Electronic Research Laboratories (SERL) at Baldock and at the Royal Signals and Radar Establishment (RSRE) at Malvern, and who have helped me to learn my 'trade' of crystal growth. Special thanks must go to my DPhil supervisors, Don Walsh and Ted Williams, to Philip Porteous, Ricky Smith and Mary Armstrong for their support during my DPhil research at RSRE, and to Phil Gurnell, Mike Rowland, Dave Wight, Norman Thompson, Derek Day, Mike Young, Verna Steward, Harry Hill, Peter Wright, David Marshall and Alison Williams for their advice and support during my years at SERL and RSRE.

In the preparation of this book, I gratefully acknowledge the help of Mary Hodgkiss and Shirley Parkinson with the typing, and Sue Beaumont with the drawings. For the information used in tables 1.2 and 1.3, my thanks go to Tom Foxon (Philips Research Labs, Redhill), David Greene (STL, Harlow), Peter Houston (University of Sheffield). and Ralph Holtom (EEV, Chelmsford). The helpful advice and encouragement of the Commissioning and Series Editors Jim Revill and Cyril Hilsum must also be acknowledged. Figures 1.3, 1.4, 1.5, 1.8, 2.6, 4.14 and 4.15 are reproduced with the permission of the Controller of Her Britannic Majesty's Stationery Office.

Finally, my sincere thanks go to my long-suffering wife Sue and sons Timothy and Christopher for tolerating long periods of neglect and allowing the dining room table to disappear under mountains of paper for hours on end.

Chapter 1

Introduction to Basic Concepts of
LPE

In this book on the application of liquid-phase epitaxy (LPE) to
the growth of III–V compounds and alloys for devices, it is
hoped to present an up-to-date assessment of the state of the art
of this technique of epitaxial deposition in terms of (i) its
applicability to various types of electronic devices, (ii) its
capabilities in comparison with other epitaxial techniques, e.g.
metal-organic vapour-phase epitaxy (MOVPE) or molecular beam
epitaxy (MBE) and (iii) the current areas of interest in LPE
research which are likely to lead to improved material quality or
novel device structures.

To this end, the book starts with some basic definitions and a
description of the important processes involved in LPE growth.
Also in Chapter 1, a comparison of LPE with MOVPE and MBE is
presented in terms of technical and operational parameters of
importance to the materials grown in a research or production
environment.

Chapter 2 presents a short history of the various techniques of
carrying out LPE growth of III–V materials, concentrating on
their relative advantages and disadvantages.

The theoretical understanding of various facets of the LPE
growth process is quite good. In Chapter 3, the use of diffusion-
limited growth theory in describing the kinetics of LPE and the
value of theoretical phase-diagram calculations for predicting the
liquid–solid equilibrium conditions for LPE are described. Also
considered are the various theories of impurity incorporation,
and the mechanisms of lattice misfit accommodation in epitaxial
structures.

In Chapter 4, firstly, the important materials parameters and
structural characteristics of device materials are considered with

examples of typical values of these parameters for several III–V materials grown by LPE. Then in Chapter 5, recent developments in several aspects of LPE growth of importance to devices are described, and a selection of examples of device applications (microwave and optoelectronic) are given which underline the areas where LPE growth can offer benefits.

Finally, in Chapter 6, a look at likely future developments in LPE technology is presented, concentrating on those aspects which could lead to either improved performance of devices or improved yields.

While the book covers the whole range of III–V compounds and alloys (ternary and quaternary), obviously we shall concentrate on the more widely used device materials such as GaAs, GaAlAs, GaP, InP, GaInAs and GaInAsP. It is apt that this book on LPE concentrates on the III–V compounds since it is in this area that LPE has made the largest impact. The expansion of LPE research has gone hand in hand with the increasing interest in the III–V compounds. We begin by briefly discussing some of the basic concepts in LPE.

1.1 Epitaxy

Although the deposition of thin films goes back to the last century, the concept of 'epitaxy' was first presented by Royer [1] in 1928. The term 'epitaxy' can be defined as 'the deposition of a single crystal layer on a single crystal substrate in such a way that the crystalline structure of the substrate is continued into the layer'. In this section, we shall use the term 'homoepitaxy' where the layer and substrate are of the same material, e.g. GaAs on GaAs, and 'heteroepitaxy' where the layer is of a different material to the substrate, e.g. $Ga_{1-x}Al_xAs$ on GaAs. The term 'liquid-phase epitaxy' is normally understood to mean the growth of epitaxial layers from solutions at elevated temperatures. It is difficult to establish the origins of the technique, but the work of Nelson [2] in 1961 is generally accepted to have been seminal.

From the definition of epitaxy given above, it follows that the two conditions below must be satisfied:

(i) that the crystal structures of layer and substrate should have the same crystallographic space group;

(ii) that the unit cell dimensions, or lattice parameter, of the layer and substrate should be closely matched, though how closely is not theoretically specified. However, as a general rule it is found that if the lattice mismatch, ε, defined as

$$\varepsilon = \frac{a_L - a_S}{a_{av}}$$

where a_L is the lattice parameter of the layer, a_S is the lattice parameter of the substrate and

$$a_{av} = \frac{a_L + a_S}{2}$$

has a value $\varepsilon \lesssim 10^{-3}$ then epitaxial growth will occur albeit with some distortion of the unit cell of the epitaxial layer in order to maintain exact lattice plane continuity across the substrate–layer interface. For $\varepsilon > 10^{-3}$, there is an increasing tendency towards the generation of misfit dislocations either at the interface or as threading dislocations, and an increasing difficulty in nucleating the epitaxial layer growth.

Using the techniques of molecular beam epitaxy (MBE) and metal-organic vapour-phase epitaxy (MOVPE) where higher supersaturations are possible than in LPE, it is possible to grow single crystal films on substrates with high lattice mismatches ($\varepsilon > 10^{-1}$) or with different crystal structures, e.g. GaAs on sapphire or GaAs on Si. In these cases, the orientation of the layer is clearly influenced by the substrate although Royer's original definition of epitaxy whereby the crystal structure of the substrate is propagated into the layer is not strictly obeyed in these cases.

1.2 Phase Equilibria

The basis of liquid-phase epitaxy is the control of the liquid–solid phase equilibrium. The principles involved may be understood by reference to figure 1.1 which shows the phase diagram for the binary system Ga–As [3]. The phase rule states that the number of degrees of freedom in a system, f, is given by

$f = c - p + 2$, where c is the number of components and p is the number of phases.

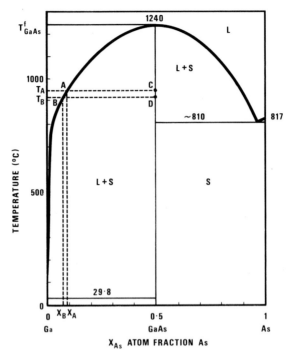

Figure 1.1 Liquid–solid phase diagram for the binary Ga–As system (after Hall [3])

For the normal conditions of LPE growth, the partial pressures of the constituents are negligibly small, and the working pressure is close to 1 atm. The pressure variable is thus neglected, and $f = c - p + 1$. For a *binary* system, e.g. Ga–As, $c = 2$ and $p = 2$ (solid, liquid), and thus $f = 1$. For a *ternary* system, e.g. Ga–Al–As, $c = 3$ and thus $f = 2$. For a *quaternary* system, e.g. Ga–Al–As–Sb, $c = 4$ and thus $f = 3$.

In figure 1.1, the upper curve represents the so-called 'liquidus'. A point on this curve, e.g. A, represents the temperature (T_A) at which a mixture of GaAs and Ga of composition X_A, if heated from lower temperatures passes from being a two-phase mixture (Ga + GaAs) to being single phase (Ga + As liquid), i.e. the point at which the last solid dissolves. When the solution is cooled through T_A, solid should begin to appear at this temperature in the absence of supercooling effects. In reality, supercooling causes there to be a range of 10–20 °C below T_A before solid material is homogeneously nucleated in the solution. The composition of the solid phase is given by the 'tie-line' AC in figure 1.1, i.e. GaAs in this case. If the temperature is dropped further to T_B, the equilibrium liquid phase composition is now X_B and the solid composition is given by 'tie-line' BD, i.e. GaAs again. If the cooling process is carried out with a GaAs seed or substrate present in the solution then epitaxial deposition of GaAs will occur on the substrate. This deposition process is called liquid-phase epitaxy (LPE). The growth solutions in III–V LPE are normally metal-rich, and the group III (metal) atom, e.g. Ga or In, is regarded as the 'solvent' and the group V atom, e.g. P, As, as the 'solute', although for convenience the III–V compound itself, e.g. GaAs, InP, is used as the solute species. Note that LPE growth can occur in the temperature range just below the liquidus temperature where the solution is 'single phase' or when there is excess solute present in the solution, i.e. 'two-phase' growth.

Although the solid phase composition in figure 1.1 is represented by a thin vertical line at $X = 0.5$, in fact the existence region has a finite width as shown in figure 1.2, showing that GaAs may exist with deviations from exact stoichiometry on both the As- and Ga-rich side [4]. It can be seen that at the low temperatures at which LPE occurs, the deviations from stoichiometry are small, whereas at the melting point, T_f, at which melt-grown crystals are produced, the deviations are much larger.

The phase diagrams for ternary and quaternary systems are more complex than for a binary system such as that in figure 1.1, and will be discussed in more detail in §3.3. The important point which applies to the LPE growth of all materials is that provided the necessary phase data is known, a particular compound or alloy can be grown over a wide range of temperatures. The factors which affect the choice of growth temperature are discussed in the next section.

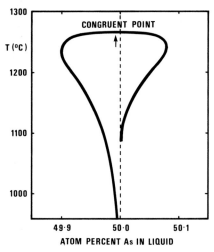

Figure 1.2 Solidus existence region for GaAs (after Hurle [4]).

1.3 Choice of LPE Growth Temperature

Although in theory LPE growth can occur anywhere along the liquidus curve, in practice there are several factors which limit the choice of growth temperature.

High growth temperatures can lead to:

(i) High epitaxial growth rates, due to the large value of $1/m$ where m is the slope of the liquidus curve in $°C$ (atom fraction)$^{-1}$. This is undesirable if very thin ($< 1 \ \mu$m) layers are to be reproducibly grown.

(ii) A higher risk of contamination of the growth solution from the container material (normally graphite, silica or boron nitride).

(iii) An increased likelihood of thermal degradation of the substrate before growth due to preferential loss of the group V constituent. This is particularly a problem with InP substrates which are used for the growth of InP, GaInAs and GaInAsP.

(iv) An increased possibility of post-growth strain being introduced during cooling to room temperature after heteroepitaxial

LPE growth due to differences of expansion coefficient between the layer and the substrate.

(v) Higher vapour pressures of solution components and dopants leading to volatilisation from the growth solution during the heating cycle prior to initiation of the layer growth. This makes control of composition and doping difficult.

Low growth temperatures can often lead to:

(i) Poor layer nucleation due to surface oxides on the substrate not being removed by reduction by the ambient gas, normally Pd-diffused hydrogen.

(ii) Low growth rates due to low values of $1/m$. This may cause difficulties if thick (> 20 μm) layers are required.

(iii) Failure to remove unwanted volatile impurities from the growth solution. This can lead to high background impurity concentrations and low mobilities.

The growth temperature chosen for a particular application is normally a compromise between these various factors, and depends on the type of device structure required in terms of layer thicknesses, carrier concentrations and other parameters.

1.4 Basic LPE Growth Processes

The LPE growth of an epitaxial layer can be broken down into several processes, any of which could in principle be the rate-determining step:

(i) Nucleation—the formation of stable clusters of atoms (called nuclei) on the substrate surface as the result of absorption, desorption and surface mobility of solute species;

(ii) Surface attachment—the stage at which nuclei grow and coalesce to produce the macroscopic growth features of the layer;

(iii) Solutal transport—the process by which atoms move towards or away from the solid/liquid interface during growth, either by diffusion or convection.

These aspects will be discussed in more detail in the following sections.

1.4.1 Nucleation and surface attachment

The theoretical treatment of the nucleation and growth of
epitaxial films has been the subject of many publications [5–7]
and will not be discussed in detail here. Faust and John [8] have
classified solution growth of III–Vs in terms of growth on (i)
singular atomically flat surfaces (111) or (100), (ii) so-called
'rough' or non-singular surfaces. The main difference is that on
rough faces, atom attachment can occur singly, and the growth
rate r would be expected to be given by $r \propto \Delta G_{cryst}$ where
ΔG_{cryst} is the change of free energy per mole when crystallisation occurs, and can be related to the solution supersaturation
by $\Delta G_{cryst} \sim RT\delta$ where δ is the supersaturation of the growth
solution, defined by

$$\delta = \frac{C(T) - C_L(T)}{C_L(T)}$$

where $C(T)$ is the solute concentration in the growth solution at
temperature T and $C_L(T)$ is the liquidus (or equilibrium) solute
concentration for the temperature T.

On the singular faces, however, atoms add in clusters and the
growth rate becomes a superlinear function of ΔG_{cryst}, and the
exact addition modes vary with crystal orientation, i.e. (100),
(111)A or (111)B. For instance, Longo *et al* [9] have discussed
the critical nucleus size for various crystallographic orientations
in terms of a model in which an atom cluster is stable when half
of the bonds of the atoms in the cluster are made either to the
substrate surface or to other cluster atoms. These orientational
differences in interface kinetics are important in LPE growth as
they lead to varying incorporation of impurities and majority
constituents in the growing layer under certain conditions. These
orientation effects will be discussed further in more detail in
Chapter 3.

Due to the experimental difficulties of observing the primary
nucleation phase of LPE growth, little work has been done on
this in comparison with vapour or molecular beam deposition.
Instead, work has concentrated on the 'growth islands' or
'hillocks' of lateral dimensions 1–100 μm which are seen at a
stage intermediate between the formation of stable nuclei and
the coalescence to form a continuous film [10–13]. This type of

study can be helpful in relating the observed surface topography of the LPE layer to experimental parameters under the control of the crystal grower. The main parameters which are found to affect the density and morphology of growth islands in LPE are as follows.

Solution supercooling
This is normally defined in terms of temperature

$$\Delta T_s = T_L - T_G$$

where T_L is the equilibrium (liquidus) temperature for the growth solution and T_G is the actual LPE growth temperature.

Typically, values of ΔT_s up to $10\,°C$ below the liquidus temperature can be sustained in III–V growth solutions before spontaneous nucleation occurs, which relieves the supersaturation.

It appears that for both homoepitaxy [11,13] and heteroepitaxy [11,12], solution supercooling can improve the surface quality by increasing the density of island formation. This leads to earlier coalescence of growth islands, i.e. when they are quite thin, and thus produces a flat surface. If the initial island density is low, then coalescence may not occur until the islands have grown to a thickness of several microns. The greater spread of island thicknesses at this stage can lead to large surface steps if two coalescing islands are of differing heights (figure 1.3). However, it is not always easy to isolate the effect of supercooling on nucleation from other effects. For example, in the case of heteroepitaxial growth it often happens that the substrate is unstable in the growth solution, e.g. GaInP on InP [14], GaAs on InP [15], GaInSb on InSb [16] and GaInAs on InAs [17], and rapid initial coverage of the substrate surface is necessary to avoid dissolution. This problem of substrate instability has been the subject of much recent research, and is discussed further in §3.3.6.

The effects of solution supercooling on LPE surface topography can be seen from figures 1.4(*a*) and (*b*), which show the as-grown surface of a layer of $Ga_{0.3}Al_{0.7}As$ grown on a GaP (111)B substrate (*a*) without supercooling and (*b*) with an initial $3\,°C$ supercooling. The latter shows a much smoother surface

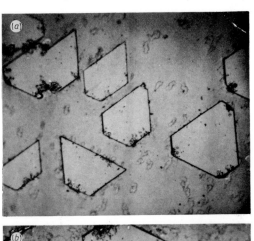

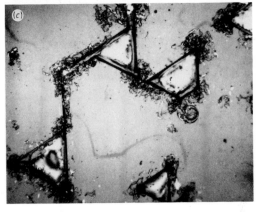

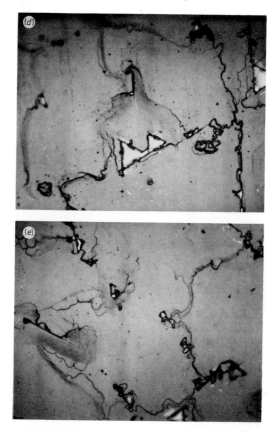

Figure 1.3 Stages in the nucleation and coalescence of islands of (GaAl)As on GaP (111) substrates grown by LPE at 800 °C (© Crown Copyright, HMSO, 1990).

due to the higher density of initial nucleation which leads to coalescence at an earlier stage where the thicknesses of the nuclei are small.

Temperature gradient at growing interface
Several workers in the early days of LPE research attributed poor LPE layer surface topography (figure 1.5) to the effects of constitutional supercooling which is commonly encountered in melt grown crystals. The phenomenon can be understood by

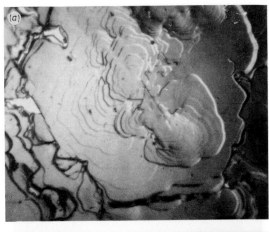

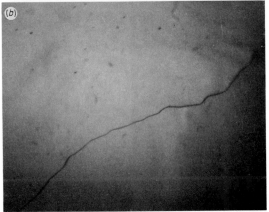

Figure 1.4 Effect of initial solution supercooling on the surface topography of LPE-grown (GaAl)As on GaP (111) substrates, with (*a*) no supercooling and (*b*) 3 °C supercooling (© Crown Copyright, HMSO, 1990).

reference to figure 1.6. During LPE growth of, for instance, GaAs from Ga-rich solutions, the solute As is depleted at the interface since its distribution coefficient $k \gg 1$ where

$$k = \frac{\text{concentration of As in solid}}{\text{concentration of As in liquid}}.$$

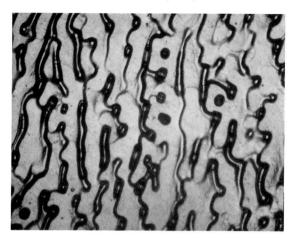

Figure 1.5 Surface of an LPE-grown InP layer showing features attributed to constitutional supercooling (© Crown Copyright, HMSO, 1990).

This depletion causes a concentration gradient in the solution ahead of the advancing interface, and this gradient is associated with a variation of effective liquidus temperature as shown by the line in figure 1.6. If a small (or zero) temperature gradient exists in the solution (line A) then the shaded region is effectively 'supercooled' below the localised liquidus temperature, and this can lead to interface instability. If a steep-enough temperature gradient (B) is imposed normal to the solid–liquid interface, the region of constitutional supercooling is removed. In LPE growth, this temperature gradient has often been achieved by cooling the substrate from below [11,18].

In this situation, it is difficult to separate the effects of the temperature gradient from supercooling effects. Mattes [18] has proposed that the presence of an interfacial temperature gradient nullifies the effect of localised thermal fluctuations caused by the latent heat of solidification and thus increases the density of nuclei. However, the design of Mattes' boat makes it likely that the solution is supersaturated at the start of growth.

Similarly, Longo *et al* [11] demonstrated for the homoepitaxial growth of GaAs that an imposed temperature gradient produced by cooling the substrate from below improved the nucleation of the epitaxial layers. Again, there is the possibility

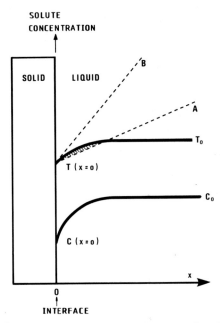

Figure 1.6 Solute concentration, C, as a function of distance x from the solid–liquid interface, and the corresponding profile of liquidus temperature. Lines A and B represent low and high interface temperature gradient respectively, with a region of constitutionally supercooled liquid.

of solution supersaturation being the major effect in this work, as pointed out by Crossley and Small [13], who suggested that the effects of constitutional supercooling should not be seen in the nucleation stage, when the concentration gradient at the interface is small.

Substrate orientation
In considering the effect of substrate orientation on the nucleation of III–V zinc-blende layers, Sangster [10] has discussed the surface atomic structure of {100}, {111} and other principal planes in the zinc-blende lattice and then considered the change in the number of 'dangling' or unsatisfied bonds on the surface when a group III or V atom bonds to the surface. If the number

of such bonds is unchanged or decreased, then there is no barrier to nucleation, whereas if the number of dangling bonds increases then nucleation is likely to be difficult on that face. For example, figure 1.7(a) shows the ideal {100} zinc-blende surface where each surface atom has two dangling bonds, while figure 1.7(b) shows the sort of structure expected of a real, rough surface, where the surface will not consist of a complete plane of A or B atoms as in the perfect case. This means that for example an A atom can add readily to the {100} surface by bonding to two B atoms, leaving two dangling bonds on the surface (i.e. no change). However, in the case of a {111} surface (figure 1.7(c)), the surface consists of a double layer of A and B atoms, wherein each A atom is bonded to three B atoms and vice versa. The next adjacent double layer is well spaced. The addition of a single A atom to a {111}B face will be by a single bond to a B surface atom, causing a net increase in dangling bonds of two. The next stage requires another A atom to bond to the surface as a second-nearest neighbour to the original one, again with a net increase of two dangling bonds. Another B atom can then bridge these two A atoms followed by a third A atom to produce the A_3B cluster shown in figure 1.7(c). Because of these considerations, it would be expected, and indeed is found experimentally, that the nucleation and growth rates on $\langle 111 \rangle$ surfaces are generally slower than for the $\langle 100 \rangle$ surfaces.

It has been widely found in III–V LPE that the surfaces have a periodic ripple or terrace structure [19–24] which is related to the misorientation of the substrate surface from a low-index plane. Figure 1.8(a) shows a typical example for InP. The terrace structure has a wavelength λ and a step height 'h' (figure 1.9). In a detailed study of the surface morphology of LPE InP, Pak *et al* [23] have concluded

(i) The terrace structure is similar in material grown in vertical dip or sliding boat systems.

(ii) Terraces appear on surfaces slightly misorientated from a low-index plane (θ in figure 1.9). This is shown in figure 1.8(b) for growth of InP on a substrate with an accurately orientated area on it.

(iii) Terrace lines are normal to the gradient of the misorientation.

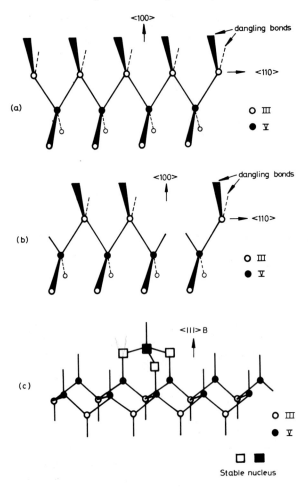

Figure 1.7 Atomic structure of zinc-blende surfaces, showing (*a*) {100} ideal surface, (*b*) real {100} rough surface and (*c*) {111} surface with a stable four-atom nucleus (after Sangster [10]).

(iv) Wavelength λ does not depend strongly on misorientation, θ, but the step height h varies with misorientation to compensate for this.

(v) Wavelength λ decreases with increasing supercooling,

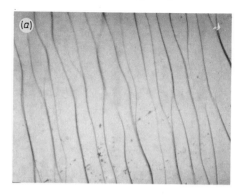

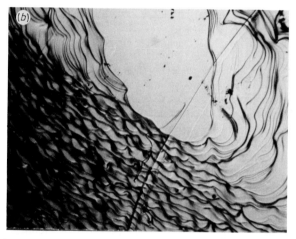

Figure 1.8 Surface photographs of LPE-grown InP layers showing (*a*) normal terrace structure due to slight misorientation of the substrate surface from a low-index plane and (*b*) terrace structure on a layer grown on a substrate with an accurately orientated ($< 0.1°$) region polished on it (upper right-hand part of photograph) (© Crown Copyright, HMSO, 1990).

probably due to the increased density of nucleation as discussed earlier.

Rode [24] introduced the concept of a characteristic riser angle φ which varies with material and orientation. If the surface misorientation θ is such that $\theta \geqslant \varphi$, then the terracing can be eliminated. However, in practice it is easier to try to remove terraces by producing a substrate surface which is exactly on orientation than with a fixed misorientation.

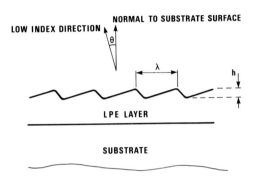

Figure 1.9 Formation mechanism of terrace on an LPE-grown layer on a substrate misorientated by θ from a low-index plane.

Substrate surface preparation
The nature of the substrate surface is a critical factor in any epitaxial growth process and can affect the LPE growth in several ways:

(i) Chemical contamination from pre-growth etching or from inside the LPE kit (O_2, H_2O) can lead to patchy or non-existent nucleation. This is a particular problem in low-temperature LPE, e.g. on InSb [25], GaSb [26] and InP [27] where native oxides may not be reduced by the ambient hydrogen. Holmes and Kamath [25] found that for low-temperature LPE growth of InSb and (Ga,In)Sb on InSb substrates the proper chemical etching of the substrate was crucial in obtaining uniform nucleation.

This effect has been exploited to achieve selected area epitaxy using Al_2O_3 [28] or SiO_2 [29] as a block to nucleation in certain areas. This will be discussed further in Chapter 5.

(ii) Decomposition of the substrate surface by preferential vaporisation of one of the constituents (e.g. P_2 from InP) can cause uneven growth with pits. This has been seen in the LPE growth of InP [30], where the surface topography is greatly improved when PH_3 is added to the ambient gas to prevent loss of phosphorus from the InP substrate.

(iii) The surface roughness of a substrate can affect the morphology of an LPE layer grown on it. For example, several authors have described the LPE growth of GaAlAs layers on roughened GaP substrates [12,31] which appears to cause a marked increase in the nucleation density and hence to smoother surfaces. However, this approach can lead to increased defect densities at the interface and in the growing layer [31], and hence has not been widely used for growth of device material. The increased nucleation density may be due either to the fact that the rough surface provides a very large surface area of high-index lattice planes on which crystal growth can readily occur by addition of single atoms, or to the fact that the roughening process produces a high density of dislocations near the surface which are known to act as nucleation sites.

Once the initial nuclei have coalesced, the LPE layer is believed to continue to grow by the lateral movement of the risers of the terraces across the layer surface by continuous surface attachment (figure 1.10). This process has been studied in an elegant series of experiments described by Bauser [32] on the LPE growth of GaAs, which showed how the macroscopic features of the layers such as impurity striations, the terrace structure and microcleavage patterns could be explained on the basis of this growth mechanism.

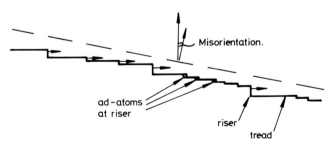

Figure 1.10 Lateral microscopic growth mechanism involving movement of risers across the surface of the epitaxial layer (after Bauser [32]).

1.4.2 Solutal diffusion

The rate-determining step in the LPE growth of III–V compounds over a wide range of temperatures and growth rates has been supposed to be the diffusion of solute species towards or away from the liquid–solid interface. The evidence for this is to some extent circumstantial in that although the diffusion-limited growth model predicts well the observed dependence of LPE growth rates on growth time and solution thickness, the absolute values of the growth rate are generally fitted by using the solute diffusion coefficient, D, as an adjustable parameter. Unfortunately, direct measurements of diffusion coefficients in typical LPE solutions have not been made, although the values normally used for fitting ($D \sim 5 \times 10^{-6}$ to 5×10^{-5} cm^2 s^{-1} depending on temperature) are typical for liquid metals.

The diffusion-limited model of LPE growth will be discussed in detail in §3.2. The fact that this model fails to predict the substrate orientation effects on growth rate implies that for many LPE growth situations it may not give a complete description of the growth kinetics.

1.5 Impurity Segregation

Each species in an LPE solution, both majority group III and V atoms and intentional and unintentional impurity atoms, is incorporated into the growing epitaxial layer to some extent, expressed by the so-called segregation or distribution coefficient, k, which is defined as

$$k = \frac{\text{concentration of species in the solid phase}}{\text{concentration of species in the liquid phase}}. \quad (1.1)$$

It is often difficult to ascertain from the values of k quoted in the literature exactly how they were calculated since different authors have used various units (often unspecified) for the dopant concentrations, e.g. atom fractions, gram atoms per unit volume or gram atoms per unit weight. In table 1.1, where segregation coefficients for several impurities in III–V materials are listed, the values have been calculated wherever possible using atom fractions, leading to the equation

$$k_i^* = \frac{(N_d - N_a)}{2N} \frac{MW_s}{\rho_s X_i^l} \quad (1.2)$$

where $(N_d - N_a)$ or $(N_a - N_d)$ is the net donor or acceptor concentration per cm^3, N is Avogadro's number, MW_s is the molecular weight of the solid, ρ_s is the density of the solid and X_i^l is the atom fraction of the impurity 'i' in the growth solution. Where this calculation has not been possible, the values are the author's original ones, and are indicated by a dagger (†). Normally, the impurity atom is electrically active as a donor or acceptor, and the carrier concentration $(N_d - N_a)$ or $(N_a - N_d)$ as measured by the Hall effect is assumed to be equal to the impurity concentration. This entails making several assumptions:

(i) The Hall factor, r, is unity.

(ii) All the impurity atoms of interest are fully ionised at room temperature.

(iii) The impurity atoms can exist only in neutral or singly ionised states.

(iv) There is negligible compensation, e.g. due to complex centres.

(v) Absence of precipitates.

In some reported work, the Hall mobility data has been analysed to calculate the values of N_d and N_a separately. Where this is the case, it has been indicated in table 1.1. Also, some authors have directly measured the chemical concentration of the dopant species, e.g. by mass spectrometry or radio-tracer techniques. Again, this type of result is indicated in table 1.1.

There are several pitfalls in comparing data from various authors on distribution coefficients. Firstly, it is important to establish that the dopant concentration in the solid is a linear function of the dopant concentration in the liquid, i.e. $[X_s] \propto [X_l]^y$ where $y \sim 1$ as shown in figure 1.11(a). Some dopant systems show the behaviour in figure 1.11(b) where at higher doping the gradient becomes ~ 0.5, and the distribution coefficient is thus concentration dependent.

In spite of these problems, the agreement between authors is fair. The high values for Zn, S and Si are a problem in LPE since these elements are commonly found in the starting materials such as Ga and In, and are electrically active in III–V materials. Although S can be readily removed by baking the growth solutions, Si is a persistent impurity which is difficult to remove.

The various theories of impurity segregation will be discussed more fully in §3.4.

Table 1.1 Distribution coefficients of dopants in LPE III–Vs.

Material	Dopant	k^*	Growth temp (°C)	Source	Orientation	Ref
GaAs	Zn	1.2×10^{-2}	800	$N_a - N_d$	(111)B	[33]
		2.5×10^{-2}	800	$N_a - N_d$	(111)A	[33]
		1.8×10^{-2}	800	$N_a - N_d$	(100)	[33]
		3×10^{-2}	880	$N_a - N_d$	(111)B	[34]
		4.5×10^{-3}	900	$N_a - N_d$	(100)/(111)B	[35]
	Ge	7×10^{-3}	900	$N_a - N_d$		[36]
		4.8×10^{-3}	900	$N_a - N_d$	(111)B	[34]
		7.5×10^{-3}	900	N_a	(100)	[34]
		1.1×10^{-4}	700	N_d	(100)	[37]
	Sn	1.2×10^{-4}	860			[38]
		7.4×10^{-5}	700	$N_d - N_a$	(100)	[39]
		2.1×10^{-4}	850	$N_d - N_a$	—	[40]
	Se	4†	800	$N_d - N_a$	(100)	[41]
		16†	690	$N_d - N_a$	(100)	[41]
		4.2	850	$N_d - N_a$	(100)	[42]
	Te	0.37	1000	Radiotracer	(111)B	[43]
		0.79	850	$N_d - N_a$	—	[40]
		1.7	770	$N_d - N_a$	—	[40]
		0.6	800	$N_d - N_a$	(111)B	[44]

Table 1.1 (*Cont.*)

Material	Dopant	k^*	Growth temp (°C)	Source	Orientation	Ref
InP	O	4.7×10^{-4}	700	$N_d - N_a$	(100)	[39]
	Cr	1.2×10^{-5}	780	Deep acceptor by DLTS	—	[45]
	Zn	0.94	650	$N_a - N_d$	(100)	[46]
		0.49	650	$N_a - N_d$	(100)	[47]
		0.73	640	$N_a - N_d$	(111)B	[48]
	Cd	1.3×10^{-3}	650	$N_a - N_d$	(111)B	[46]
		6.5×10^{-4}	670	$N_a - N_d$	(100)	[49]
		1.1×10^{-4}	570	$N_a - N_d$	(100)	[49]
	Si	2.5	650	Microprobe	(100)/(111)	[46]
		30†	550–715	$N_d - N_a$	(100)/(111)	[50]
	Ge	2.8×10^{-3}	650	$N_d - N_a$	(111)B	[46]
		5.5×10^{-3}	650	$N_d - N_a$	(111)B	[51]
	Sn	1.2×10^{-3}	650	$N_d - N_a$	(111)B	[46]
		1.2×10^{-3}	650	$N_d - N_a$	(111)B	[51]
		1.1×10^{-3}	650	$N_d - N_a$	(100)	[47]
	S	0.2–5	670	$N_d - N_a$	(111)B	[52]

Table 1.1 (*Cont.*)

Material	Dopant	k^*	Growth temp (°C)	Source	Orientation	Ref
	Se	2.1	670	$N_d - N_a$	(111)B	[52]
	Te	0.20	650	$N_d - N_a$	(111)B	[51]
		0.17	650	$N_d - N_a$	(111)B	[46]
		0.21	670	$N_d - N_a$	(111)B	[52]
	Be	0.05	650	$N_a - N_d$	(100)	[47]
GaP	Te	0.25	935	Radiotracer	—	[53]
		0.1	1135	Radiotracer	—	[53]
		0.35	900	$N_d - N_a$	(100)	[54]
	S	0.14†	1000	$N_d - N_a$	(111)B	[55]
		0.48†	1000	$N_d - N_a$	(100)	[55]
		0.17†	1000	$N_d - N_a$	(111)B	[55]
	Zn	1.3	1000	$N_d - N_a$	(111)B	[55]
		4.5×10^{-2}	1000	$N_a - N_d$	(111)B	[55]
$Ga_{0.47}-In_{0.53}-$As	Sn	2.8×10^{-3}	620	$N_d - N_a$	—	[56]
	Zn	0.63	620	$N_a - N_d$	—	[56]
GaInAsP $\lambda = 1.15\ \mu m$	Sn	4.7×10^{-3}	650	$N_d - N_a$	(100)	[57]
	Zn	0.13	650	$N_a - N_d$	(100)	[57]
	Te	0.38	650	$N_d - N_a$	(100)	[57]

†These values are the author's quoted values and are not calculated using equation (1.2).

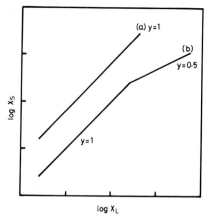

Figure 1.11 Log impurity concentration in solid (X_S) against log impurity concentration in solution (X_L), for cases of (a) linear relationship over the complete doping range and (b) change of slope at higher doping levels.

1.6 Advantages and Disadvantages of LPE as a Growth Technique for Device Materials

In Chapter 4, a more complete discussion of the capabilities of liquid-phase epitaxy for the production of device materials and structures will be presented. Here we shall examine a range of topics of relevance to the choice of a materials growth system: (i) economic and operational, and (ii) technical. To do this, the following sections will discuss the advantages and disadvantages of LPE under the broad headings mentioned above, and finally in tables 1.2 and 1.3 we shall compare LPE with molecular beam epitaxy (MBE) and metal-organic vapour-phase epitaxy (MOVPE) under a variety of headings. Because of the great diversity of LPE techniques and applications and constantly changing prices and technology, it is not meaningful to enter into great detail. The information presented here should only be considered as a general guide.

1.6.1 Economic considerations
One of the great advantages of LPE is that one can, at little expense, set up a basic system which can then be easily altered

and improved as required. This is an important consideration, particularly in research work where flexibility and adaptability are important. For example, an LPE system consisting of a home-made resistance-heated 'gold-reflector' furnace, silica work tubes with cone and socket joints, basic three-term (PID) temperature controller, a ramp generator for temperature ramping, a source of dried but non-Pd-diffused hydrogen ambient gas, together with basic peripherals such as a temperature indicator, moisture meter, hydrogen leak detector, chart recorder and gas flow meters and valves could cost about £8k in materials and be constructed in ~ 0.25 to 0.50 man years.

A more sophisticated system, incorporating a three-zone or heat-pipe furnace, a more expensive temperature controller with adjustable PID adjustments, an electronic ramp generator for controlling the temperature cooling cycles, a palladium diffuser for H_2 purification and using stainless steel gas lines with some sort of gas-tight seal to the silica work tube, could be built for £12k in materials, and take 0.5 to 0.75 man-years in construction.

There has been increasing interest in the past 10 years in the automation of the LPE process to some degree, using microcomputers or process timers to monitor and control the operation. The automation can be applied to (i) control of the temperature cycle, (ii) initiation of linear movements to either operate the slider in a sliding boat system or to move the substrate holder in a dipping system at the appropriate temperatures, (iii) control and switching of gas flows through the LPE system and the monitoring of moisture or O_2 levels in the gas stream and (iv) introduction of safety features, e.g. to switch off furnace power in the event of a failure in the H_2 gas supply.

The necessary interfaces to enable the microcomputer to drive stepper motors, actuate relays, operate mass-flow controllers, and to monitor or control temperatures using thermocouple inputs are commercially available and are reasonably cheap (£250–500 each). An automated system incorporating all the features mentioned above could be built at a cost of £30k to £80k in materials and hardware and would take about *1 to 2 man-years* to construct and commission. (Prices quoted are at 1989 levels.)

Further refinements could include:

dry box for boat loading (£3k)
vacuum system for pumping out LPE kit (£2–5k) for leak checking or flushing
high sensitivity hygrometer or O_2 meters (£2–4k)
mass spectrometer for monitoring ambient gas (£10–20k).
hydrogen leak detector (£2–4k).

1.6.2 Operational considerations

This is a more difficult area to quantify, since much depends on the way in which the system operates.

Taking into account the costs of services (electricity, gas supplies, cryogenics), repairs and maintenance, starting materials (solutions, substrates) and depreciation, a figure of $\sim$ £30k per year is obtained. However, the level of technical expertise required to run the growth system is fairly high. This is because, even with an automated system such as that described above, a considerable amount of reasonably skilled effort is required in the preparation and weighing of the starting materials and substrates and the loading of the boat. However, Small *et al* [58], for example, have described a method for precasting pellets of growth solution on a mass basis which reduces much of the preparation time. The use of precast solutions is now extremely common, reducing the level of operator skill required. In LPE, the growth solutions can normally only be used once. However, the amount of solution in each growth well is normally only 5 to 10 g, at a cost of £20–40.

The rate of production from an LPE kit in terms of wafers per day is usually lower than for a commercial MOVPE or MBE system by a factor of 2–5. This is a result of the fact that, for a sliding boat system (see §2.1.3), a solution can normally only grow one layer. To increase the output per kit, it is necessary to construct a boat which can accommodate several sets of solutions in parallel. Alternatively, larger-area substrates can be used, although this will make the furnace required bigger. For dip systems (see §2.1.2), it is possible to scale the solution up to several kilograms and to grow on several large-area substrates at once, although it is difficult to extend this to multilayer growth, since the transfer from one solution to another is more cumbersome than in the sliding boat systems.

The yield of devices per LPE slice may sometimes be reduced by incomplete solution wipe-off and/or uneven grown surfaces. The development of fully automated LPE systems has led to improvements in the output of slices per kit and in the yield of device-quality material, as well as the control of layer thickness in the < 1 μm range.

An advantage of LPE systems is their versatility, in that a wide variety of materials and/or device structures can be grown in a growth apparatus without major modifications. This can be a distinct advantage in a research environment where the materials requirements may continually change. Changes of materials, e.g. from GaAs to InP, often require a change in growth temperature leading to adjustments to furnace temperature profiles. Large alterations to layer thickness can cause problems in that a different LPE growth technique or different boat designs might be required.

In summary, an LPE system is relatively cheap to build, maintain and run, but the rate of production is low in comparison with other epitaxy methods. This is partly due to the long times required to achieve a stable situation prior to the start of growth. While this is not an insuperable problem, there are relatively few papers in the literature addressing this particular area. Another important operational consideration is safety. Apart from the use of H_2 gas as an inert ambient in the LPE work tube, which necessitates careful elimination of O_2 from the growth system, there are no notable hazards in the LPE growth of III–V compounds, since the starting materials are reasonably safe when handled with tweezers, and the vapour pressures of the solution constituents during growth are low.

A comparison of several of the factors discussed above for LPE, MBE and MOVPE is shown in table 1.2.

1.6.3 Technical factors

The main reason for the initial interest in LPE growth of III–V materials was the prospect of high efficiency LEDs, lasers and detectors operating in the visible and near infrared ($\lambda = 0.9$– 2 μm), although much of the early work on GaAs Gunn-effect microwave devices also used LPE-grown material.

Table 1.2 Comparison of operational factors in epitaxial growth of III–Vs by LPE, MBE and MOVPE.

	LPE	Comments	MBE	Comments	MOVPE	Comments
Cost of equipment £k (1989) (research)	30		600		180	Not including laboratory safety equipment
(production)	80		350		250	
Cost of operation per year per production equipment £k (1989) (incl. power, maintenance, raw materials, depreciation)	30		120	Mostly depreciation	80	
Operator skill level required 1 (low) to 5 (high)	4	Can be reduced by use of automated systems and precast solutions	5/1	A highly skilled operator is needed to oversee some aspects e.g. reloading cells, leak checking with unskilled operators working under him	4	
Man-hours per growth run	4		1–3		1–3	

Table 1.2 (*Cont.*)

		LPE	Comments	MBE	Comments	MOVPE	Comments
Growth runs per day	(research)	1		1–3		2–3	
	(production)	2		2–3		2–3	
No of substrates per growth run	(research)	1		1		1–2	
	(production)	1		7		5–6	
Size of substrates (mm)	(research)	10 × 10		50 diam.		20 diam.	
	(production)	30 × 20		75 diam.		50 diam.	
Degree of automation		High	Both research and production equipment is now commonly fully automated	High	Not all production equipment fully automated, but can be readily achieved	High	Nearly all production equipment is fully automated. The process lends itself to automatic control

Table 1.2 (*Cont.*)

	LPE	Comments	MBE	Comments	MOVPE	Comments
Versatility of growth equipment (scale of 1 (low) to 5 (high))	4	Wide range of materials can be grown in same equipment	3	Many materials can be grown in one equipment; some memory effects from group V elements. Difficult to grow alloy with two group V species	3	Same basic equipment can be used for all III–V compounds if required alkyls and hydrides are plumbed in. However, InP-related alloys require special equipment
Adaptability of growth process—scale 1 (low) to 5 (high)	3	Adjustments of furnace profiles may be required and trial runs to optimise growth temperatures	4	Several weeks' work on calibration runs for a new material are required	4	Optimum growth conditions need to be established for new material; these are dependent on reactor geometry
Safety hazards—scale of process—scale 1 (minimal) to 5 (severe)	2	H_2 ambient gas commonly used. Care in loading and leak checking required	1	Slight risk during maintenance and loading of Knudsen cells	5	Alkyls are often pyrophoric; hydrides highly toxic; large quantities of H_2 used. Rigorous training required

The advantages that LPE had to offer were:

(i) high luminescence efficiency due to the low concentration of non-radiative centres and deep levels;
(ii) growth of ternary and quaternary alloys to greatly extend the range of materials properties available;
(iii) controlled n-type and p-type doping with a wide variety of dopants available;
(iv) ability to grow multilayer structures, e.g. p–n junctions, heterostructures with low interface recombination velocities;
(v) good reproducibility and uniformity of materials properties over large areas.

These advantages have been exploited in a wide range of optoelectronic devices, particularly those where the device is of small area so that many devices can be made from a single slice. The aspects in which LPE is not so advantageous are where:

(i) large areas (>2 cm^2) of material are required to be free of surface blemishes (e.g. for transmission photocathodes or integrated circuits);
(ii) very abrupt or precisely controlled doping or composition profiles are required (e.g. for Gunn or IMPATT diodes);
(iii) extremely accurate layer thickness uniformity and reproducibility are required (e.g. for microwave FETS and quantum-well structures);
(iv) compositional grading to overcome lattice mismatch between layer and substrate is desired.

To fulfil these requirements, the Ga(In)-trihalide-H$_2$ and more recently the metal-organic VPE techniques have been developed and much research devoted to the technique of molecular beam epitaxy (MBE). It is in the field of microwave devices such as Gunn diodes, microwave FETS and IMPATT diodes and in the new area of multiple-quantum-well devices, such as MQW lasers and high electron mobility transistors (HEMT) where the device performance is critically dependent on the abruptness of interfaces and the uniformity and control of epitaxial layer thickness, that LPE has found little application although some authors have reported successful LPE growth of microwave devices such as GaAs mm-wave Gunn oscillators [59], GaAs

Table 1.3 Comparison of materials quality parameters between LPE, MBE and MOVPE grown III–V epilayers.

Factor	LPE	Comments	MBE	Comments	MOVPE	Comments
Run-to-run reproducibility of layer thickness	± 10%		± 5%		± 5%	
Uniformity of layer thickness (across 20 mm)	± 5% (best) ± 10% (typ.)	Edge growth effects occur	± 5%		± 2% (best) ± 5% (typ.)	Edge growth effects occur
Run-to-run reproducibility of alloy composition, Δx (e.g. in $Ga_{1-x}Al_xAs$)	± 1%		< ± 1%		± 1%	
Uniformity of alloy composition, Δx (across 20 mm)	± 1%		< ± 1%		< ± 1%	
Depth uniformity of alloy composition (1–10 μm layer thickness)	~ 2%	Depends on segregation coefficient and solution depletion effects	< 1%		~ 1%	

Table 1.3 (*Cont.*)

Factor	LPE	Comments	MBE	Comments	MOVPE	Comments
Electron mobilities at 77 K $(cm^2 V^{-1} s^{-1})$ — GaAs $\{$	8×10^4 2×10^5	Typical Best	1×10^5 1.5×10^5	Typical Best	5×10^4 1.5×10^5	Typical Best
InP $\{$	4×10^4 1×10^5	Typical Best	—		4×10^4 1.4×10^5	
Deep level concentration (cm^{-3}) in GaAs	$< 10^{12}$ $< 10^{13}$	Electron traps Hole traps	$< 10^{11}$ $< 10^{11}$	Electron traps Hole traps	$< 10^{13}$ $< 10^{12}$	Electron traps Hole traps
Interface abruptness (10–90%) in GaAs–(GaAl)As (Å)	~ 100 ~ 500	Best Typical	~ 3 ~ 15	Monolayer, best Typical	3–100	Dependent on growth apparatus
Alloy or dopant grading capability (on scale 1 = difficult, 5 = readily achieved)	1	Can only exploit depletion effects of high segregation coefficients	4		5	Good control achieved by ramping flow rates of metal-organics
Surface perfection; density of surface blemishes (hillocks, pits, pinholes) (cm^{-2})	~ 10	Main feature is ripple pattern related to substrate misorientation, amplitude 0.01–0.1 μm	~ 500	Related to C-contamination on substrate surface and to substrate dislocations	~ 200	Dependent on substrate misorientation and III/V ratio in gas phase

Table 1.3 (*Cont.*)

Factor	LPE	Comments	MBE	Comments	MOVPE	Comments
Depth uniformity of carrier concentration ($1-10$ μm layer thickness)	10%		20%		10%	
Background carrier concentration at 77 K (cm^{-3}) — GaAs	4×10^{12} (best) 10^{15} (typ.)	Very dependent on solution baking procedures	$< 10^{13}$ (best) 10^{14} (typ.)		Low 10^{14} (best) 10^{15} (typ.)	Dependent on purity of alkyls and hydride starting materials
— InP	10^{14} (best) 10^{15} (typ.)				Low 10^{14} (best) 10^{15} (typ.)	
Doping capabilities for GaAs (cm^{-3})	n $\{$ Up to 2×10^{19}	Sn dopant	Up to 3×10^{18}		Up to 10^{19}	Se dopant
	p $\{$ Up to 2×10^{19}	Ge dopant	Up to 3×10^{19}	Be dopant	Up to 10^{20}	Zn dopant
Run-to-run control of carrier concentration (GaAs)	$\pm 20\%$		$\pm 20\%$		$\pm 10\%$	

MESFETS [60] and GaAs IMPATT diodes [59] and multiple-quantum-well devices such as GaInAsP MQW lasers at $\lambda = 1.3\ \mu\text{m}$ [61]. Current advances in LPE techniques are enabling improved device material or structures to be grown. Some examples are:

(i) controlled vapour pressure during LPE growth to reduce defect density in layers;

(ii) substrate etch-back before growth to produce improved interface perfection and better surface topography;

(iii) selected-area epitaxy and growth on structured substrates to produce novel heterostructure laser devices;

(iv) automation of LPE systems to improve reproducibility of layer thickness for short growth times (< 1 s).

Advances such as these mean that the III–V materials grown by liquid-phase epitaxy are steadily improving in quality and some of them will be discussed in further detail in Chapter 5. However, the main challenge remains to transfer these laboratory-scale improvements into the realm of factory-scale production.

Table 1.3 presents a comparison of the technical capabilities of LPE compared to those of MBE and MOVPE under a variety of headings. Where values of materials parameters are compared, the GaAs/(GaAl)As system has been used as it is the one which has been most researched and for which plenty of data is available in the literature.

The typical properties of LPE-grown materials are discussed in more detail in Chapter 4.

Chapter 2

Short Summary of LPE Apparatus and Techniques

Over the years, the field of research into LPE has been an extremely fertile one which has produced a wealth of novel approaches in the areas of equipment design and epitaxial growth techniques. The basic requirement is to bring the substrate and growth solution into contact while the epitaxial layer grows, and then to separate them at the end of the growth cycle.

In this chapter, the various ways in which these basic requirements have been met in different equipment designs and growth techniques over the past 15 to 20 years will be summarised. Table 2.1 provides a 'star-rating' comparison of the various techniques.

2.1 LPE Growth Systems and Boat Designs

2.1.1 Tip system
This is probably the earliest type of LPE system used for the growth of III–V compounds and was first proposed by Nelson [2]. The basic design is shown in figure 2.1, and consists of a furnace set on a base which can be tilted slightly about a horizontal axis. The boat, normally made from graphite or silica, sits in the work tube in the centre of the tilted furnace such that at the start of the experiment the growth solution is at one end of the boat, and the substrate is at the other, held in place by a clamp. When the equilibration of the solution is complete, the furnace temperature is slowly reduced and the furnace tipped to roll the solution over the substrate. After a suitable time, the furnace is tipped back to its starting position,

and the solution rolls off the grown layer. This method is restricted to the growth by single layers, although it has been used for growing p–n junctions in InP by dropping a pellet of p-type dopant into the growth solution part-way through the growth run [62].

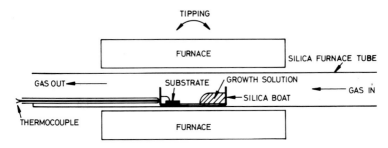

Figure 2.1 Tipping LPE furnace system (after Nelson [2]).

The attraction of the tip technique is that the apparatus is extremely simple and cheap to assemble, and it is for this reason that it was much used for doing preliminary experiments in the early days of LPE. Its main drawback, apart from the difficulty of growing multilayer structures, is that the removal of the solution at the end of the growth period is unreliable. In a development of the tipping technique, Panish *et al* [63] used a sliding graphite substrate holder which moved the solution on and off the substrate when the work tube was tipped. This technique, illustrated in figure 2.2, gives better wipe-off and was the forerunner of the sliding boat system which is now the most common method of LPE growth (see §2.1.3).

2.1.2 Dip system
The 'dip' technique was originally described by Rupprecht [64] and the configuration used is shown in figure 2.3. The growth solution is held in a crucible, usually of silica or graphite, with monitoring and control thermocouples located beneath the crucible. The substrate(s) is located either horizontally or verti-cally in a silica holder. The outer tube is made of silica and is fitted with an air-tight PTFE or O-ring coupling, which allows the

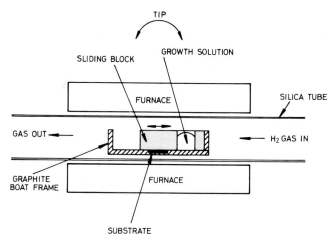

Figure 2.2 Tipping LPE system with sliding boat (after Panish *et al* [63]).

substrate holder to be moved vertically or rotated about the vertical axis. Heating is provided either by resistance or RF heating. Normally, there is a small temperature gradient through the solution with the surface hotter than the bottom of the solution. This gradient is designed to reduce convection in the solution and to avoid constitutional supercooling.

One of the advantages of this technique is that the substrate can be kept in the cool upper part of the furnace tube prior to epitaxial growth, which is important if the substrate contains a volatile constituent which may be rapidly lost from the surface at the growth temperature, e.g. in the growth of GaP on GaP substrates at temperatures in the range 900–1100 °C. Another advantage of this technique is the simplicity of the apparatus, which made the dipping system a popular one in the early days of LPE for preliminary experiments, particularly as it is possible to measure the liquidus temperature by direct observation of the solution, by having a suitable viewing port at the top of the work tube.

The major disadvantages of the dip technique are the non-uniformity of layer thickness across a slice, and the relatively large volumes of solution required. While it was found possible to re-use solutions up to 50 times [21], difficulties arose in

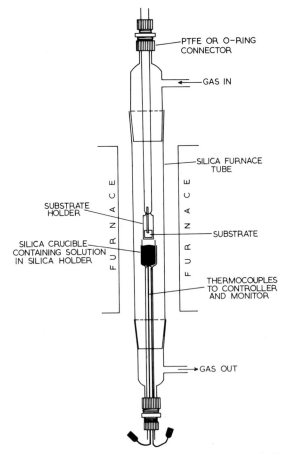

Figure 2.3 Dipping LPE system (after Rupprecht [64]).

preventing contamination of the solutions or loss of components between runs, thus making precise control of doping and growth rate difficult.

Another problem with the dipping technique is due to the fact that the substrate passes through the solution surface which can often be contaminated with oxide or other particles. This can lead to poor nucleation on the substrate surface but can be overcome by arranging for a wiper to sweep the surface of the solution clean before the substrate enters.

The complete removal of solution at the end of the growth

cycle is often not achieved, which can lead to rough surfaces and make multilayer growth difficult due to cross-contamination of the growth solutions. To overcome this problem, a rotating substrate method has been developed [65] to spin off the solution at the end of the growth period, as shown in figure 2.4. This variation of the dip technique also produced more uniform layer thicknesses.

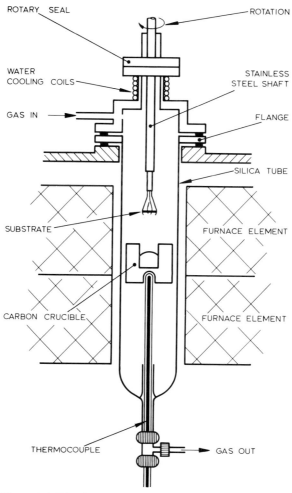

Figure 2.4 Dipping LPE system with fast substrate rotation (after Astles *et al* [65]).

In summary, the dipping system, while still used on a small scale for experimental research work, has not found widespread use in industry since it is unsuited to the growth of the more complex III–V epitaxial structures of current interest. However, a large-scale version of the dipping method, the infinite melt technique, has been developed by workers at Santa Barbara Research Center, USA [25] for growth of InSb and is currently being used to produce LPE-grown (CdHg)Te layers on a production basis. In this case the large growth solutions (several kg in weight) are re-used many times since the amount of solute removed in each run is proportionally very small.

2.1.3 Horizontal sliding boat system

It is true to say that the horizontal sliding boat system of LPE has proved to be the most popular and versatile of the various growth techniques. The origin of the method is difficult to establish, but certainly the paper by Panish *et al* [66] was one of the first. Figure 2.5 shows the fundamental features of the technique while figure 2.6 shows a photograph of the component parts of a boat assembly. The substrate normally sits in a recess in the base of the boat while the growth solutions reside in wells machined in the movable sliding block. The solutions may be positioned over the substrate as required by moving the sliding block using an external push-rod arrangement.

The boat material is usually graphite which has several advantages:

(i) it can be easily machined into the necessary shapes to high tolerances;
(ii) it is available in high-purity form;
(iii) it is self-lubricating;
(iv) it is inert to all the normal solution constituents;
(v) it has a high thermal conductivity;
(vi) it is not wetted by the growth solutions.

However, it does have the disadvantage of being porous, which leads to absorption of atmospheric O_2 and H_2O during loading, but this can be overcome by the use of a dry-box for loading the boats, and by careful baking procedures.

Other materials such as silica [67,68], glassy carbon [69] and sapphire [70] have been used, and have yielded high-purity LPE

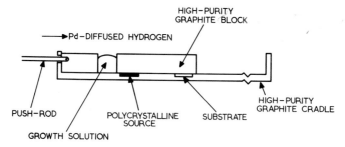

Figure 2.5 Basic structure of graphite horizontal sliding boat (after Panish *et al* [66]).

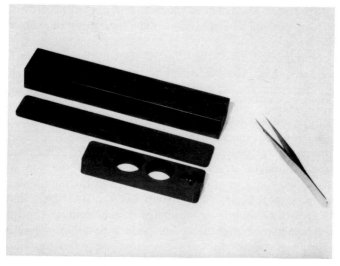

Figure 2.6 Main component parts of sliding boat assembly: (from top) the outer frame, the base plate incorporating a recess for the substrate, and the slider incorporating two circular solution wells (© Crown Copyright, HMSO, 1988).

layers, but have not found widespread use due to the difficulty of manufacture.

Empty wells or slots are sometimes provided between full wells in the sliding block to help in wiping off solution left on the substrate. A later innovation was the use of lids and weighted pistons on top of the solutions, the purpose of these being respectively to prevent the escape of volatile species from the solution and to ensure that the solution both completely fills the corners of the well and has a constant thickness.

An important factor in the boat design is the depth of the substrate surface below the surrounding graphite surface. This wiping clearance is important; if it is too large solution wipe-off will not be efficient, while if it is too small scratching of the substrate may occur during the slider movements. Typical clearances are in the range 25–100 μm (1–4 thou″). Most systems are arranged so that the substrate stays in position while the block containing the solutions is moved so that each solution in turn is in registration with the substrate, the dwell time being determined by the layer thickness which is required to be grown. In other boat designs, it is the base plate containing the substrate which moves.

A common feature of this type of boat is the provision of a source slice which, like the substrate, sits in a recess in the base plate. This may consist of a slice of polycrystalline or single crystal material which can be used to presaturate a solution or to remove excess supersaturation from a solution prior to it being moved onto the substrate.

More complex boats have a large number of solutions for growing multilayer structures, entailing several solution wells and wiper slots and several source positions in the base plate.

The horizontal sliding boat system has had its biggest impact in the growth of multilayer structures where accurate control of layer thicknesses is important and good wipe-off of solution at each stage is important to avoid cross-contamination of the solutions. Because the volume of the solution is normally much less than in tip and dip systems, economical use is made of the solution materials. For most applications, the depths of solution used are in the range 1 to 10 mm and typical solution well dimensions are from $\sim$ 1 cm^2 (e.g. 10 mm $\times$ 10 mm square cross section) to $\sim$ 6 cm^2 with typical solution weights being in the range 0.5 to 10 g.

The thin solutions typically used in sliding boat systems have the added advantage that effects due to either thermal or solutal convection are minimised. The Rayleigh number R for a liquid is the factor which gives a measure of the tendency for thermal or solutal convection to take place and is proportional to d^3 where d is the depth of fluid. It is thus advantageous to reduce the solution depth as much as possible to reduce the likelihood of convective disturbance. Thin solutions also reduce the problem of excess edge growth which is sometimes encountered in LPE growth and can be particularly troublesome in sliding boat systems causing surface scratching during slider movements. This problem will be discussed further in §4.5.3.

Some exceedingly ingenious designs have been developed around the basic sliding boat principle. Two of these will be discussed in further detail in the following sections, namely the 'rotary slider' and the 'wipingless' boat system.

2.1.4 Rotary slider

The rotary slider technique was a development of the linear slider technique described in the previous section. Figure 2.7 shows the design of Thompson and Kirkby [71] which consists of a graphite cylinder with several small wells machined in it to contain the growth solutions. The substrate and any source discs are located in the circular base plate. The solutions are moved from position to position by rotation of the cylinder about the

Figure 2.7 Rotary slider boat components (after Thompson and Kirkby [71]). (Reproduced by permission of North-Holland Publishing Company.)

vertical axis. This type of boat has been almost exclusively used for the growth of double-heterostructure lasers where the requirement for a four- or five-layer structure on fairly small area substrates is ideally matched to the capabilities of this system. A particular advantage is the good temperature uniformity that can be achieved around the circumference of the rotary crucible, which helps to achieve reproducible growth of thin layers.

2.1.5 'Wipingless' growth system

One of the difficulties sometimes encountered with the horizontal sliding boat system, particularly when multilayer structures are being grown, is the problem of scratching of the epitaxial layer surface during the slider movements. This is often caused by the dragging of small crystallinities heterogeneously nucleated in the solution across the surface of the layer. Another problem which can arise when the growth ambient contains undesirable impurities such as H_2O or oxygen is that the surface of the epitaxial layers may be contaminated during the short period of time that the slice is uncovered during the sliding movements. To overcome these difficulties, the 'wipingless' method was developed by Horikoshi [72] and is shown in figure 2.8. For full details of the boat operation the reader is referred to Horikoshi's paper [72], but the basic principle is that each growth solution is forced by a piston through an input slit into a chamber containing the substrate. At the end of the growth of each layer, the new solution entering through the input slit replaces the old one which is forced out through the output slit into a reservoir. Although complex, this boat design worked well and was capable of reproducibly growing GaAs/(GaAl)As double-heterostructure lasers. There appeared to be no dramatic difference between the material grown in this type of boat compared to the normal sliding boat system. Since the problems of slice scratching and interface contamination in the standard slider boat can be eliminated by careful boat design, there does not appear to be clear advantage in the use of this type of boat. There are also two possible disadvantages with the system. Firstly, there is a period during the interchange of solutions when a mixed solution of indeterminate liquidus temperature may be produced leading to either etch-back or spontaneous growth. Secondly, a large proportion of the growth solution is wasted, in that it is not actually used in the growth of the epitaxial layer, but is flushed to waste.

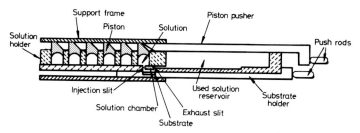

Figure 2.8 'Wipingless' horizontal sliding boat system (after Horikoshi [72]).

2.1.6 Multi-slice boat designs

There have been several papers published on the subject of carrying out simultaneous LPE growth on many substrates at once, with the aim of increasing the output of wafers per growth run [73–75]. The first approach was to deposit single layers onto many substrates [73]. However, more recently, boats have been designed which allow multilayer growths. For example, Heinen [74] has reported a boat which is capable of growing a four-layer structure onto 16 substrates simultaneously, although in the trial experiments only a three-layer (GaAl)As structure onto GaAs was produced. The principle of the design is shown in figure 2.9, with the substrates being held vertically in pairs on either side of graphite strips, which can move as shown to transport the wafers successively through the chambers containing the growth solutions. These chambers are filled from upper reservoirs (not shown) after equilibration of the growth solutions. The results from this system in terms of surface perfection, uniformity of alloy composition and uniformity of thickness were similar to those obtained on conventional single substrate systems. Dutt *et al* [75] have improved the design of Heinen [74] to allow the growth of InGaAsP and InGaAs alloys on InP. The main changes are (i) to allow the solutions to drop from the upper reservoirs into the growth solution chambers at different times and (ii) to provide for the generation of a phosphorus overpressure in the slider to protect the InP substrates. Modification (i) permits the use of single-phase growth solutions (i.e. no excess solid present) by allowing the solutions to remain in contact with a source piece of InP until they are needed in the growth sequences, while (ii) ensures that no loss of phosphorus occurs from the InP substrate surface. Dutt *et al* [75] found that this system gave good wipe-off of solution, smooth surface topography and good slice-to-slice uniformity of alloy composition, although slice-to-slice layer thickness varied by ± 20%.

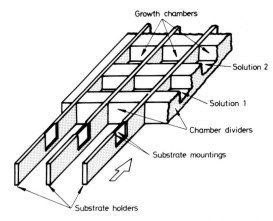

Figure 2.9 Multisubstrate LPE boat design allowing growth of up to four layers on up to 16 substrates (after Heinen [74]).

In neither of these papers [74,75] was the full 16-slice capability tested and it remains to be seen whether these designs will prove practicable in a production environment. However, they illustrate the scope for ingenuity in the basic graphite sliding boat design.

2.2 Ancillary LPE Equipment

Apart from the boat in which the growth is carried out, there are some other important aspects of an LPE system which merit brief discussion:

- furnace design
- growth ambient
- vacuum attachments
- automatic control systems

Since the design and construction of an LPE system, like any other piece of scientific apparatus, is subject to individual preference and prejudice, many of the comments which follow are bound to reflect the author's own experience.

2.2.1 Furnace design
Due to the differing requirements of various research laboratories, it is unusual to purchase a commercial furnace specifically designed for LPE work. Most laboratories construct their own furnaces for their specific purposes. The main considerations of a furnace design for LPE work are as follows.

(i) *Thermal mass of the furnace.* This may be large or small depending on the type of growth being performed. High thermal mass furnaces are generally employed where temperature stability is important, e.g. where the surface quality or growth rate of the layer might be affected by oscillations in the furnace temperature. This can be achieved by employing bulky insulation between the heating element and the outer furnace walls or by using a heavy furnace liner such as a mullite or alumina tube. More recently, heat pipe furnace liners usually made from stainless steel have become widely used to improve the flatness of the furnace temperature profile. These also have very large thermal masses and thus improve the temperature stability of the furnace.

Low thermal mass furnaces are useful where rapid temperature changes are required, for example when supercooled or step-cooled growth is being performed. The familiar 'gold' furnace [76] has become popular for this purpose, since it also allows good visibility inside the work tube. The main item in this design is the outer gold reflector tube which consists of a silica tube coated on the inside with a thin, semitransparent film of gold, surrounding the silica former on which the heating element is located. By reflecting most of the IR radiation back into the tube, acceptable temperature stability and good temperature profiles may be obtained. These furnaces are now commercially available.

(ii) *Temperature profile*. For most LPE work it is important for the temperature profile of the furnace to have a reasonably long region in the centre where the temperature is constant to within ±1 °C, the length of this 'flat zone' depending on the dimensions of the LPE boat being used. Occasionally, a profile with a temperature gradient may be required. In either case, it is important that the profile is reproducible from day-to-day. This makes it undesirable to use heating units in which the elements are likely to move or to degrade during use. The windings can be held in place on the former by helical notches or grooves, or encased in a thermal cement.

The use of axial or concentric heat pipes in furnace construction has become more common in the past few years, and commercial designs are available to cover the range 200–1200 °C. Heat pipes have an extremely high effective thermal conductivity, and are thus used to produce a constant temperature zone in the furnace to accommodate the growth boat. Heat pipes have an evacuated space containing a working vapour, normally Hg (200–400 °C), K (400–1000 °C) or Na (500–1100 °C) in equilibrium with the liquid held on a gauze. Since the vapour pressure is constant throughout the volume of the heat pipe, the temperature must be constant. This follows from a consideration of the phase rule $f = c - p + 2$ (see §1.2). Here the number of components $c = 1$ and the number of phases $p = 2$, and thus the degrees of freedom $f = 1$. Flat zones can also be achieved by using three-zone furnaces with the ends being adjusted to give a central constant temperature zone or to produce non-symmetrical profiles where these are required. The use of heat-pipes simplifies the furnace construction and reduces the amount of control equipment by removing the need for three zones. It also improves the temperature stability as mentioned in the previous section.

(iii) *Visibility*. If visibility of the LPE boat is important, e.g. for performing *in situ* liquidus measurements, then by far the best furnace design is the 'gold'-coated reflector type mentioned earlier. It is possible to cut observation ports in more conventional furnaces but care must be taken not to affect the temperature profile.

2.2.2 Ambient gas
The ambient gas inside the work tube is normally palladium-

diffused hydrogen produced by commercially available units (Mattheson, Johnson–Matthey). This gas is of high purity, normally containing < 1 PPM by volume of H_2O and O_2. Some workers, considering the possibility of a reaction at high temperatures between $SiO_2(s)$ and $H_2(g)$ to produce $SiO(g)$ leading to silicon contamination of the growth solution, have used gases such as Ar or N_2 dried by molecular sieve and deoxygenated by passing over a heated catalyst. However, hydrogen has the advantage of being able to reduce any metal oxides which may be present on the surface of the growth solutions and is hence the most commonly used gas.

The effect of small concentrations of O_2 or H_2O in the ambient gas during LPE growth on the purity and morphology [69,77] of epitaxial layers makes the measurement of the moisture content of the ambient gas during growth experiments extremely important. Several commercially available hygrometers are suitable for this purpose. Care must be taken to filter out any volatile material which might be carried into the hygrometer by the output gas flow. Solid state sensors are now available which are more sensitive, have faster response and are easier to calibrate.

Some workers monitor the O_2 level in the output gas using a Hersch cell or other detector, though this may not be a reliable measure of the oxygen content of the incoming hydrogen gas, since at the normal LPE temperatures the equilibrium

$$H_2(g) + \tfrac{1}{2}O_2(g) \rightleftarrows H_2O(g)$$

will be well over to the right-hand side.

The control of the partial pressure of the more volatile (group V) constituent in the vapour phase using a separate source, e.g. $P_4(g)$ in the growth of GaP, has been shown to be an important factor affecting the crystalline perfection of the growing layer [80,81]. However, this aspect of LPE growth of III–Vs has been little investigated and merits further research.

2.2.3 Vacuum attachments
In recent years, it has become common to find vacuum systems attached to LPE kits. At their most basic these may consist of a small roughening pump (which may be a rotary or sorption pump) used to evacuate the LPE work tube after loading, followed by backfilling with nitrogen or hydrogen. A more

sophisticated arrangement may consist of a high-vacuum system consisting of turbo-molecular, ion, diffusion or titanium sublimation pumps for thorough outgassing and leak checking, possibly incorporating a mass spectrometer to monitor the presence of impurity gases in the growth environment. The vacuum system can also be used for leak testing the growth apparatus or for *in situ* vacuum baking of the growth boat.

2.2.4 Automatic control systems

As the demands placed on the epitaxial crystal grower by the device engineer have become more exacting over the years in terms of (i) control of layer thickness, doping etc and (ii) the number of layers in a device structure coupled with the transfer of LPE and other epitaxial techniques into production, there has been an increasing interest in computer-controlled LPE equipment. Since there is, to the author's knowledge, no commercial supplier of complete systems, each laboratory or research organisation has designed and built up its own growth system around a commercially available computer. It is thus impossible to generalise about the actual equipment used, but the basic components are normally:

(i) a microcomputer with peripherals such as a floppy disk drive for storage of programs and experimental data, a printer for hard copy of data and programs, VDU for data displays and an *X-Y* plotter for producing parameter plots, e.g. temperature versus time;

(ii) an interface system which can both turn the commands given by the computer into signals that operate external pieces of equipment and also receive data from sensors on the growth system, and which inputs them to the computer where the signal can be compared to the demanded value thus completing the control loop;

(iii) external units and sensors which are under the control of the computer or supply data to the computer. These would normally include: temperature controllers and ramp generators for controlling the furnaces; mass flow controllers and meters for gas handling; stepper motors and drivers to provide the linear or rotary movements for the sliding boats; a vacuum system for leak checking and roughening out; hygrometers or oxygen meters for measuring the purity of the ambient gas in the LPE apparatus; potentiometric or optical sensors to indicate the positions of furnaces and push-rods.

Although the development of the computer software and the commissioning of such a system can be a complex and lengthy process, the improvements in reproducibility and productivity which derive from LPE automation make the investment well worthwhile.

2.3 LPE Growth Techniques

The basis of LPE growth is the production of supersaturation in the growth solution, such that the deposition of solid material occurs onto the substrate. This supersaturation can be produced in several ways and this section will discuss briefly each of the following (see figure 2.10):

(i) *Ramp-cooled growth (RC)* where the temperature of the solution is lowered at a rate R (°C min^{-1}) from the liquidus temperature (T_L) to a temperature $T_L - \Delta T_R$ whilst in contact with the substrate (figure 2.10(i)).

(ii) *Two-phase solution growth (TPS)* which is a variation of RC growth where solid material is present in the solution throughout the growth cycle, due either to the addition of excess solute material or to having cooled the solution so far below the liquidus temperature that spontaneous nucleation has occurred.

(iii) *Step-cooled growth (SC)* where the solution is held at a temperature ΔT_s below the liquidus temperature and then brought into contact with the substrate and held at constant temperature during the growth cycle (figure 2.10(ii)).

(iv) *Supercooled growth (S)* where the temperature of the growth solution is lowered ΔT_s below the liquidus temperature (T_L). The substrate and solution are then brought into contact while the solution is cooled at a rate R (°C min^{-1}) over a range ΔT_R (figure 2.10(iii)).

(v) *Constant-temperature-gradient growth (CTG)* where the driving force for growth is produced by a temperature difference across the growth solution, with a source and substrate placed in

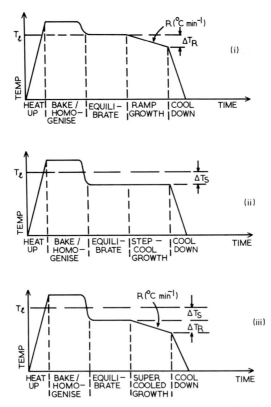

Figure 2.10 Temperature against time profiles for (i) ramp-cooled, (ii) step-cooled, and (iii) supercooled LPE growth methods.

the solution, such that the source is at a higher temperature than the substrate. Solute material is transported from the source to the seed.

(vi) *Transient-mode (TMLE) growth* in which a cool substrate is introduced into a hotter solution which can be either saturated or supersaturated.

The growth technique chosen depends on the material parameters which are of importance for the particular application. The chief materials features which are affected by the choice of growth technique are epitaxial layer thickness control, uniformity and reproducibility of composition for ternary and quaternary alloys, and surface topography. In the following sections, the various growth techniques will be briefly discussed with these features in mind.

2.3.1 Ramp-cooled growth

Figure 2.11 shows calculated lines of layer thickness (d) versus time (t) obtained using the equations to be derived in Chapter 3 for the case of ramp-cooled growth of GaAs at 800 °C [82]. These follow the relationship $d \propto t^{3/2}$, and experimentally this time-dependence is followed quite closely. By comparison with the lines for step-cooled growth shown in figure 2.12, it can be seen that by suitable choice of cooling rate R, a slow growth rate can be obtained just as readily by this technique as by step-cooled growth. In practice, however, the ramp-cooled growth technique does not give such good results, with the thickness uniformity and surface topography tending to be inferior to that obtained in step-cooled growth [82]. This is particularly true in the case of heteroepitaxial growth when there is a lattice mismatch between the layer and substrate. Also, experimental growth conditions are more difficult to reproduce from run to run in ramp-cooled growth, and the

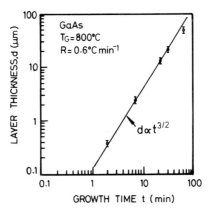

Figure 2.11 Layer thickness d (μm) against growth time t (min) for ramp-cooled LPE growth of GaAs at 800 °C (after Hsieh [82]).

growth is more sensitive to temperature fluctuations in the furnace. In spite of these difficulties, this growth technique has been widely used particularly when thicker layers (5–50 μm) are required or where the layer thickness is not critical.

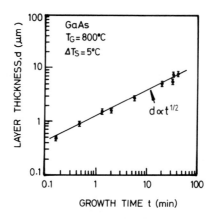

Figure 2.12 Layer thickness d (μm) against growth time t (min) for step-cooled LPE growth of GaAs at 800 °C (after Hsieh [82]).

There have been recent reports [83] that in the LPE growth of lattice-matched structures of GaInAsP and InP for $\lambda = 1.35\ \mu$m double-heterostructure lasers, better laser performance is obtained when the layers are grown by slow ramp-cooling cycles, the explanation being that there is less alloy composition grading in the thin quaternary active laser layer due to the slower growth rate achieved by the ramp-cooling technique.

2.3.2 Two-phase solution growth
This technique is similar to the ramp-cool variant except for the presence of excess solid material in the solution during the cooling cycle. This has the advantage in several situations of removing the necessity for accurately weighing small quantities of saturating materials into the solution, e.g. in the growth of GaInAsP on InP [84,85], but has the disadvantage that the growth rates can be variable due to the presence of competing nucleation sites in the solution. Also, if two solid materials are

added, e.g. GaAs and InP to In + Ga mixtures to produce GaInAsP growth solutions, there is the danger that variations in the rate of dissolution of the two solid materials can mean that slightly different points on the liquidus surface might be reached in each run, thus leading to variations in solid composition.

2.3.3 Step-cooled growth

It was mentioned in §2.3.1 that while both the ramp-cool and step-cool growth techniques can give low growth rates when thin layers are required, the latter technique has two main advantages:

(i) the supercooling present in the growth solution enhances nucleation, which can be important when there is a lattice mismatch between the epitaxial layer and the substrate;

(ii) the step-cool technique is a constant temperature process, which can eliminate problems due to the temperature-dependence of the layer composition [86] or of impurity incorporation.

Theory predicts that in this growth method d should vary as $d \propto t^{1/2}$ and this relationship has been found to hold experimentally (figure 2.12). This method has been widely used in the LPE growth of III–V compounds, most commonly in lattice-matched heteroepitaxial growth, e.g. (GaIn)(AsP) on InP [86–88], (GaIn)As on InP [89], (GaIn)(AsSb) on GaSb [90].

The main limitation of this method is the fact that the growth solution will only support a limited supercooling (normally about 10–15 °C for metal-rich III–V solutions) before spontaneous crystallisation occurs and relieves the supercooling in the solution. Thus, when thicker layers ($> 10 \, \mu m$) are needed with good surface flatness, the supercooled growth technique is used.

2.3.4 Supercooled growth

This technique is really a hybrid, in that the early stages of growth are basically a step-cooled regime whilst the ramp-cooling becomes more dominant in the later stages.

This growth technique has the advantage of an initial supercooling to enhance nucleation together with the ability to grow thicker layers than can be obtained by simple step-cooled growth. It has been widely used in the heteroepitaxial growth of several III–V alloys: (GaIn)As on InP [91–93], Ga(As,Sb) on GaAs [94], (GaIn)(AsP) on InP [95–97], (GaIn)(AsSb) on GaSb [98,99].

2.3.5 Temperature-gradient growth

This type of growth has not been as widely used as the others, due to the more difficult boat and furnace design. The various geometries which have been employed are shown in figure 2.13, with (*a*) and (*b*) showing a vertical geometry with a source and substrate wafer held in a temperature gradient at constant temperature in the growth solution either (*a*) with the source above the substrate or (*b*) with the source below the substrate. Some workers have used a temperature gradient in conjunction with a normal ramp-cooling, figure 2.13(*c*), to overcome constitutional supercooling effects. Figure 2.14 shows a diagram of a silica holder used by the author in a vertical dipping system in the mode of figure 2.13(*b*) for the growth of thick layers of GaP and GaAs.

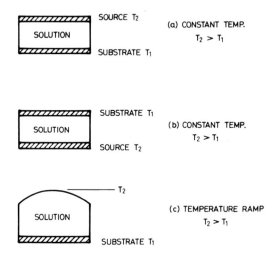

Figure 2.13 Various geometries of constant-temperature-gradient LPE growth with (*a*) the solution being hotter at the top, (*b*) the solution being hotter at the base and (*c*) the solution hotter at the top, with no source and a cooling ramp applied.

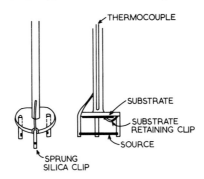

Figure 2.14 Holder for carrying out constant-temperature-gradient LPE growth in the geometry of figure 2.13(*b*).

The temperature gradient can be imposed in several ways:

(i) by running the upper and lower parts of a horizontal-axis furnace at different temperatures [42,100];

(ii) by utilising the natural temperature gradients at the top or bottom regions of a vertical axis furnace [100,101];

(iii) by adding an auxiliary heater above the solution well in a sliding boat system [81];

(iv) by cooling the lower part of the solution using a flow of gas through a hollow compartment in the base of the graphite boat, thus cooling the substrate from below [18].

This technique has several advantages:

(i) the presence of the temperature gradient can help to avoid constitutional supercooling [18,102];

(ii) the epitaxial layer grows at constant temperature, thus avoiding the changes in alloy composition or impurity incorporation which occur when the growth occurs over a range of temperature [81];

(iii) in the arrangements shown in figure 2.13(*a*) and (*b*), the solute is continuously replenished by the source during the growth cycle, thus enabling thick (> 100 μm) layers to be grown.

2.3.6 Transient-mode growth

This technique was first used by Deitch [103] for the homoepitaxial growth of GaAs, and involved dipping a cool substrate into a saturated solution in a vertical dipping system. The initial growth rate and the nucleation density are both high, which can be extremely favourable when growing highly lattice-mismatched structures.

In more recent work, Moon and van der Plas [15] have used a horizontal sliding boat system placed in a non-uniform temperature profile such that the substrate is held at a temperature $\sim 30\,^{\circ}\mathrm{C}$ lower than the solution prior to contact. This technique has been successfully applied to the growth of highly lattice-mismatched structures, e.g. GaAs on InP ($\Delta a/a \sim 3.6\%$) and (GaAl)As on GaP ($\Delta a/a \sim 3.8\%$).

While the control of layer thickness is difficult with this technique, it may be the only practical way for growing certain structures where there is a large lattice mismatch or a problem with the stability of the substrate in the growth solution.

2.3.7 Electroepitaxy (current-controlled LPE)
The technique of electroepitaxy in which an electric current is passed through the liquid–solid interface to stimulate layer growth has been the subject of much research [104–110]. The major effect was initially believed to be Peltier cooling at the metal (solution)–semiconductor (solid) interface although it is now believed that the observed growth rates are too high to be explained by the Peltier effect alone, and electromigration is thought to be the dominant mechanism. The effect had initially been investigated by Pfann *et al* [111] in relation to zone melting of germanium. The first reported application in LPE was in the growth of InSb [105], where a direct relationship between growth rate and current density was found. Figure 2.15 shows a diagram of the boat used in this work which is typical of that used by other workers. The design is based on the normal horizontal graphite sliding boat, but in this case, the top and bottom boat parts are electrically isolated from each other in order that the current flow occurs only through the solid–liquid interface. The substrate sits on a graphite pedestal although later workers have used a GaAs pedestal in order to reduce the effective thermal conductivity of the substrate, K_s. This is important since the Peltier cooling ΔT_p is given by

$$\Delta T_p = PI/(1 + K_1/K_2)K_s + K_1 \qquad (2.1)$$

where K_1, K_2 and K_s are the thermal conductivities of the growth solution, the base on which the substrate sits, and the substrate respectively, I is the current and P the Peltier coefficient.

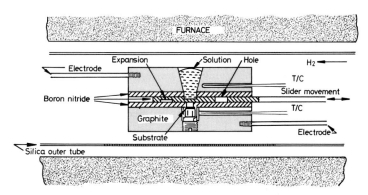

Figure 2.15 Boat used for liquid-phase electroepitaxial growth of InSb (after Kumagawa [105]).

Kumagawa [105] grew layers both by current control only, and by superimposing current pulses on a normal ramp-cooling LPE growth. In the latter case, it was demonstrated that the dopant incorporation was affected by the current pulses leading to striations which showed up by chemical etching. This ability to 'mark' the layer growth has been used to monitor the morphology and growth rate of epitaxial layers [108].

The characteristics of epitaxial layers grown by electroepitaxy and those grown by normal 'thermal' LPE have been compared by Imamura [107,109]. In particular, the characteristic surface ripple is not seen in layers grown by electroepitaxy [109] and it has been reported that the density of certain microdefects was much lower in layers grown by electroepitaxy, while the minority carrier lifetime was increased [107].

The effect of current density on dopant incorporation has been investigated, e.g. Te in GaAs [104], and showed that reproducible increases in doping level could be obtained by increasing the current density during the growth. An even more interesting effect was found in the growth of $Ga_{1-x}Al_xAs$ layers on GaAs substrates, where an extremely uniform alloy composition, x, was obtained in thick layers grown by electroepitaxy [106] in contrast to layers grown by the normal LPE technique which have composition gradients caused by the high distribution coefficient of Al. This work has been extended to grow mm-thick layers [112] of $Ga_{1-x}Al_xAs$ with uniform x. It can thus be seen that the technique of electroepitaxy has yielded some very interesting results which have shed light on some of the processes occurring in LPE growth solutions. It is a field which is obviously deserving of more research effort in the future.

2.4 Summary

The preceding sections have shown the wide range of approaches to LPE growth and it is true to say that almost every laboratory worldwide has developed its own very individual designs of apparatus. This is partly due to the diversity of device structures and materials grown by LPE and partly due to the fact that there has never been a commercially produced LPE growth system so that researchers have been forced to develop their own.

It is difficult to compare the capabilities of these various growth techniques. An attempt at this has been made in table 2.1, looking at such factors as economy, ability to grow multilayers, efficiency of wipe-off, surface quality, thickness control and ability to scale up. It can be seen that the various types of sliding boat system come out well in most respects, but that the others have particular features which make them useful for certain applications.

Table 2.1 Comparison of the capabilities of the various LPE techniques. (☆ poor; ☆☆ average; ☆☆☆ good.)

Technique	Economy of solution usage (single run)	Multilayer growth	Solution removal	Surface quality	Thickness control	Scale up	Major advantages
Dipping	☆	☆☆	☆☆	☆☆	☆	☆☆☆	Good for growth of thick layers. Substrate can be kept cool prior to growth. Economy can be improved by using solution several times
Tipping	☆☆	☆	☆	☆☆	☆☆	☆	Good for initial growth experiments. Boat design simple. Easily adapted to sealed tube experiments.
Horizontal slider	☆☆☆	☆☆☆	☆☆☆	☆☆☆	☆☆☆	☆☆	Ideal for multilayer growths and for growth of thin layers.
Rotary slider	☆☆☆	☆☆☆	☆☆☆	☆☆☆	☆☆☆	☆☆	Designed for growth of multilayer structures—equipment is more compact than horizontal slider system.
Wipingless slider	☆	☆☆☆	☆☆☆	☆☆☆	☆☆	☆	Avoids scratching and wipe-off problems sometimes encountered with horizontal slider.

Chapter 3

Theory

3.1 Introduction

One of the major factors in the development of the LPE technique has been the successful application of theoretical calculations and models to explain and predict various aspects of the LPE growth process. In this chapter, three particularly important areas will be discussed:

1. Diffusion-limited growth theory
2. Theoretical solid–liquid phase diagram calculations
3. Impurity incorporation models.

Certain shortcomings of these theories will be mentioned, in particular the difficulty in explaining 'orientation' effects on growth rate, impurity incorporation and alloy composition, by which is meant the different values of these properties obtained on different substrate orientations.

3.2 Diffusion-limited Growth Theory

This theory has been widely used to account for LPE growth kinetics for a wide range of growth temperatures and growth techniques, e.g. step-cooling, ramp-cooling [88,113–116].

3.2.1 Basic equations
The basis of the theory is that the rate-limiting step in the growth process is the diffusion of solute species to the growth interface, i.e. that interface kinetics are fast. For the growth of a binary compound, e.g. GaAs, the one-dimensional diffusion equations are fairly simple:

$$\frac{\partial C}{\partial t} = D \frac{\partial^2 C}{\partial x^2} + v \frac{\partial C}{\partial x} \qquad (3.1)$$

for solute diffusion in liquid and solid phases, and

$$\frac{\partial T}{\partial t} = K \frac{\partial^2 T}{\partial x^2} + v \frac{\partial T}{\partial x} \tag{3.2}$$

for heat diffusion in liquid and solid phases, where C is the solute (e.g. As) concentration, and is a function of growth time, t, and distance x from the interface in a direction normal to the solid–liquid phase boundary, as shown in figure 3.1. D is the diffusion coefficient of the solute species in the solvent (e.g. As in Ga), T is the temperature, v the growth velocity and K the thermal diffusivity.

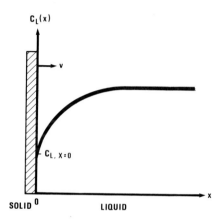

Figure 3.1 Solute concentration profile in LPE solution.

In addition, there are continuity equations describing the fluxes of heat and solute atoms at the interface

$$v(C_{S,x=0} - C_{L,x=0}) = D_L \partial C_L/\partial x|_{x=0} - D_S \partial C_S/\partial x|_{x=0} \tag{3.3}$$

$$vL\rho_S = K_S \partial T_S/\partial x|_{x=0} - K_L \partial T_L/\partial x|_{x=0} \tag{3.4}$$

where K_S, K_L are the thermal conductivities of solid and liquid phase, L is the latent heat diffusion, and ρ_S is the density of the solid phase.

For a ternary or quaternary solution, e.g. Ga + Al + As or Ga + In + As + P, the equations become more complex since the components of the growth solution may not diffuse independently, and there will therefore be terms of the form $D_{ij}\partial C_j/\partial x$

describing the flux of atom i due to the concentration gradient of species j. Due to this complexity, these terms have been taken as zero [117]. We shall merely consider the case of the growth of a binary compound. There are several simplifying assumptions that can be made for most conditions of LPE growth. These are as follows:

(i) The growth solution is isothermal—this is true for most LPE situations except those where a temperature gradient is deliberately imposed. This means that equations (3.2) and (3.4) may be neglected.

(ii) Diffusion in the solid phase is negligibly slow, i.e. $D_S \ll D_L$—this is a good approximation for normal LPE conditions of temperature and growth rate, which means that only equation (3.1) for the liquid phase and equation (3.3) (neglecting the second term on the right-hand side) need be solved.

(iii) The growth velocity v is low enough that the term $v\partial C/\partial x$ in equation (3.1) is negligible. This assumption has been shown to be an extremely good one by earlier workers [115,118], for typical LPE growth rates of 10^{-6}–10^{-5} cm s^{-1}.

(iv) The solution depth l is small enough that the effects of solutal or thermal convection may be disregarded. The tendency of a solution to become unstable due to density inversion caused either by solute or thermal gradients can be characterised by the relevant Rayleigh number

$$N_{\text{R,thermal}} = g\Delta T\alpha l^3/Kv \qquad (3.5)$$

$$N_{\text{R,solute}} = g\Delta C\gamma l^3/Dv \qquad (3.6)$$

where g is the acceleration due to gravity,
            ΔT the temperature difference across the solution,
            ΔC the concentration difference,
            α the coefficient of volume expansion,
            γ the change of solution volume per unit change of solute concentration,
            D the mass diffusivity,
            K the thermal diffusivity and
            v is the kinematic viscosity.

Cellular convection occurs when N_R exceeds a critical value of ~1000. Long [100] has calculated the critical solution height, l_{crit}, for the onset of thermal convection in a Ga–As solution under typical LPE conditions using the known physical parameters for liquid Ga. For a typical LPE solution depth of 2–3 mm in a sliding boat system, the vertical temperature gradient would have to be $>100\,°C\,cm^{-1}$ to initiate thermal convection.

For solutal convection, Tiller [102] has estimated that for LPE growth of GaP under normal conditions, no convection should be obtained for a solution depth $l < 2\,mm$. In sliding boat systems, the solution is normally isothermal with the substrate at the bottom of the solution. This arrangement leads to a stabilising solutal gradient since for most III–V systems the density of the solute is less than the density of the solvent, and thus the region of the solution near the substrate surface which is depleted of solute will have a higher density than the upper region of the solution (see figure 3.2).

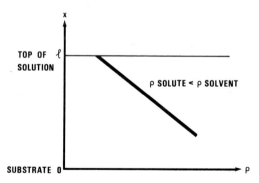

Figure 3.2 Density variation in LPE solution.

3.2.2 Solution of the diffusion equations

The analysis of the growth process with these assumptions becomes a problem in solving the equation

$$\frac{\partial C(x,\,t)}{\partial t} = D\,\frac{\partial^2 C(x,\,t)}{\partial x^2} \qquad (3.7)$$

subject to the relevant boundary conditions. The thickness of the grown layer at time t is given by

$$d(t) = \frac{1}{C_S} \int_0^t D\left(\frac{\partial C}{\partial x}\right)_{x=0} dt. \qquad (3.8)$$

The basic boundary conditions are:

(i) The growth solution is isothermal (except for temperature-gradient growth).

(ii) The solute concentration is initially homogeneous, i.e. $C(x, t = 0) = C_0$ for all x.

(iii) The solute concentration at the solid–liquid interface at any time t, $C(x = 0, t)$, has a value equal to the liquidus (equilibrium) concentration corresponding to the solution temperature 'T' at that time, $C_E(T)$, i.e.

$$C(x = 0, t) = C_E(T) \qquad \text{for all } t. \qquad (3.9)$$

The temperature T will be a function of growth time t which depends on the growth technique being considered.

(iv) The dimensions of the growth solution affect the boundary condition at the solution boundary remote from the growth interface. The following two conditions are normally specified.

(*a*) *Semi-infinite solution* where during the growth period 't' the solute concentration gradient does not reach the solution surface, i.e.

$$\left(\frac{\partial C}{\partial x}\right)_{x \to \infty} = 0 \qquad \text{for all } t > 0 \qquad (3.10)$$

(see figure 3.3(*a*)). This is a good approximation if $t < l^2/D$ where l is the solution depth. For a Ga solution at 800 °C with $D \sim 4 \times 10^5$ cm^2 s^{-1} [116] and a depth $l = 2$ mm, the condition is $t < 1000$ s (17 min), which is generally true for LPE growth.

(*b*) *Bounded solution growth* when $t > l^2/D$, the value of l becomes an important parameter since the surface concentration of solute becomes a time-varying function (figure 3.3(*b*)). For this case, the boundary condition is $\partial C/\partial x = 0$ at $x = l$.

Since the 'semi-infinite' approximation is a good one for most LPE growth cycles, only this case will be considered in the discussion which follows.

(v) In the case of source material floating on the solution surface, either in the form of excess lumps of solute or a polycrystalline source disk, then the solute profile will be as shown in figure 3.3(*c*), and the boundary conditions are

$$C(x = 0, t) = C_E(T)$$
$$C(x = L, t) = C_E(T) \qquad (3.11)$$
$$\left(\frac{\partial C}{\partial x}\right)_{x=L/2} = 0.$$

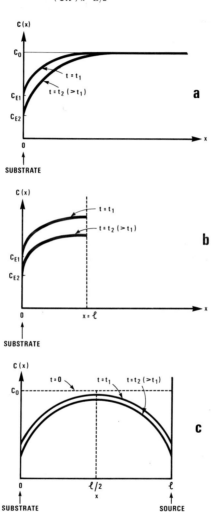

Figure 3.3 Solute profiles for (*a*) semi-infinite growth, (*b*) bounded solution and (*c*) source and seed arrangement.

Other boundary conditions, dependent on the growth technique used, are discussed in the following sections.

3.2.3 Results for various techniques

Ramp-cooled growth
Here, the solution is initially at the liquidus temperature, T_0, with a uniform solute concentration C_0, and so

$$C(x, t = 0) = C_0 \qquad \text{for all } x. \qquad (3.12)$$

The temperature is reduced linearly with time, at a rate α; such that

$$T(t) = T_0 - \alpha t. \qquad (3.13)$$

If it is assumed that for small cooling intervals ($<10\,°C$) the liquidus composition is a linear function of temperature, i.e.

$$\frac{dT_E}{dC_E} = m$$

then

$$C(x = 0, t) = C_0 - \alpha t/m. \qquad (3.14)$$

The solution to the diffusion equation for range-cooled growth over a small temperature interval using the semi-infinite solution approximation has been derived by Small and Barnes [114] to be

$$d(t) = \frac{4}{3}\left(\frac{D}{\pi}\right)^{1/2}\left(\frac{\alpha}{C_0(1 - k)m}\right)t^{3/2} \qquad (3.15)$$

where $d(t)$ is the epitaxial layer thickness at time t, k is the solute segregation coefficient at the solid–liquid interface, i.e. $k = C_S/C_0$.

If the further assumption is made that $k \ll 1$, then (3.15) becomes

$$d(t) = \frac{4}{3}\left(\frac{D}{\pi}\right)^{1/2}\left(\frac{\alpha}{C_0 m}\right)t^{3/2}. \qquad (3.16)$$

This is the equation derived by Hsieh [82], and the good fit of this equation to the experimental results on ramp-cooled growth of GaAs was shown earlier (figure 2.11).

Step-cooled growth

In this technique, the LPE solution initially has a uniform solute concentration C_0 corresponding to a liquidus temperature T_0. The solution is then cooled to a temperature $T_0 - \Delta T_S$ and placed in contact with the substrate

$$C(x = 0, t) = C_0 - \frac{\Delta T_S}{m} \qquad \text{for all } t \qquad (3.17)$$

where ΔT_S is small enough that the liquidus line can be considered linear.

The solution of the diffusion equation for this case has been presented by Rode [116], Ghez [119] and Hsieh [82]. The latter derived the equation

$$d(t) = 2\Delta T_S \left(\frac{D}{\pi}\right)^{1/2} \left(\frac{1}{C_S m}\right) t^{1/2}. \qquad (3.18)$$

The validity of this $t^{1/2}$ dependence has been demonstrated for the step-cooled growth of several materials. Figure 3.4 shows the results for GaInAsP from the work of Cook *et al* [120]. Note that the experimental points deviate from the $t^{1/2}$ relationships for $t > 30$ min, which probably indicates the limit of applicability of the 'semi-infinite' approximation.

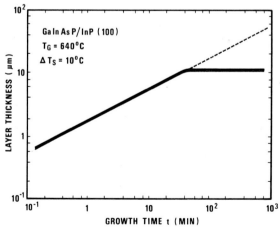

Figure 3.4 Layer thickness (μm) against time (min) for step-cooled LPE growth of GaInAsP on InP (100) (after Cook *et al* [120]).

Supercooled growth May apply

The starting conditions here are similar to those for step-cooled growth, with the solution being supercooled by ΔT_S.

However, before the solution is contacted with the substrate, a cooling ramp of α is initiated and maintained throughout the growth cycle. Therefore

$$C(x = 0, t) = C_0 - \frac{\Delta T_S}{m} - \frac{\alpha t}{m}. \qquad (3.19)$$

The solution to the diffusion equation for this case has been presented by Hsieh [82] and, not surprisingly, is a combination of 'step-cool' and 'ramp-cool' terms:

$$d(t) = \frac{1}{C_S m} \left(\frac{D}{\pi}\right)^{1/2} \left(2\Delta T_S t^{1/2} + \frac{4}{3} R t^{3/2}\right). \qquad (3.20)$$

The experimental results of Hsieh [82] are shown in figure 3.5. It can be seen that at low t they fall close to the theoretical line for step-cooled growth, while at higher t the results become closer to the ramp-cool theory. In between, equation (3.20) provides a very good description of the growth-rate behaviour.

Constant-temperature-gradient growth – Steady state – don't incl,

The solution of the diffusion equation (3.1) for this growth method has been discussed by Long *et al* [100]. The $\partial C/\partial t$ term is now zero, and the equation to be solved is thus

$$D \frac{\partial^2 C(x)}{\partial^2 x} + v \frac{\partial C(x)}{\partial x} = 0. \qquad (3.21)$$

The boundary conditions are

$$\left. \begin{array}{l} C(x = 0) = C_E(T_0) \\[2mm] C(x = l) = C_E(T_1) \end{array} \right\} \text{ for all } t \qquad (3.22)$$

as shown in figure 3.6. $C_E(T)$ is the equilibrium solute concentration at temperature T. The solution presented by Long *et al* [100] is

$$v = \frac{acD(T_0 - T_1)}{lT_0^2[a - C_S \exp(-b + c/T_0)]} \qquad (3.23)$$

where v is the growth rate, C_S is the solute concentration in the solid, D is the diffusion coefficient and a, b, c are coefficients in the expression for the liquidus curve in the form

$$C_E(T) = a \exp(b - c/T) \qquad (3.24)$$

e.g. for GaAs, $a = 4.82 \times 10^{22}$, $b = 8.42$ and $c = 1.32 \times 10^4$ [121].

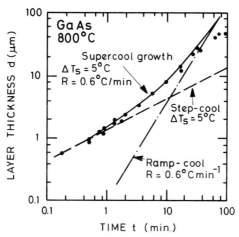

Figure 3.5 Layer thickness (μm) against time (min) for supercooled LPE growth of GaAs at 800 °C (after Hsieh [82]).

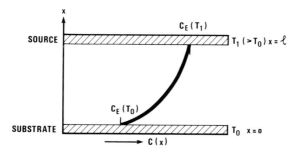

Figure 3.6 Solute profile during constant-temperature-gradient LPE growth.

Equation (3.24) is applicable to small values of $(T_0 - T_1)$, and shows that the growth rate is independent of time and directly proportional to the temperature gradient $(T_0 - T_1)/l$. Long *et al* [100] showed that this equation fitted the experimental results well if the source-seed geometry was designed to reduce the effects of solutal convection which otherwise dominates the growth behaviour.

3.2.4 Orientation effects
The fact that differences in growth rate are observed on different crystallographic orientations [65,82,123–125], means that the diffusion-limited model does not give a complete description of the growth process. The equations presented earlier are correct only if the interface kinetics are fast. The effect of interface kinetics has been considered by Bolkhovity-anov and Zembatov [126] for the case of step-cooled growth of GaAs and (Ga, In)As. Bolkhovityanov and Zembatov [126] assumed that the growth rate v (limited by interface kinetics) is given by

$$v = B(C(x = 0) - C_0) \qquad (3.25)$$

where $C(x = 0)$ is the interface solute concentration, C_0 is the equilibrium solute concentration at the growth temperature T_0 and B is the so-called kinetic coefficient.

Since $vC_S = D(\partial C/\partial x)_{x=0}$, where C_S is the solute concentration in the solid phase, then

$$\left(\frac{\partial C}{\partial x}\right)_{x=0} = \frac{BC_S}{D}(C(x = 0) - C_0). \qquad (3.26)$$

The normal diffusion equation

$$D\frac{\partial^2 C}{\partial x^2} = \frac{\partial C}{\partial t}$$

can be solved using the usual boundary conditions and (3.26), the result being

$$d(t) = \frac{2\Delta T}{C_S m}\left(\frac{D}{\pi}\right)^{1/2} - \frac{\Delta T}{fC_S m}\left[1 - \exp\left(\frac{f^2}{Dt}\right)\mathrm{erfc}\left(\frac{f}{\sqrt{Dt}}\right)\right]$$

$$(3.27)$$

where ΔT is the initial supercooling in the solution, and

$$f = BC_S/D.$$

The second term in equation (3.27) is additional to Hsieh's equation for step-cooled growth (equation (3.18)), and takes account of the interface kinetics. As $B \to \infty$, so the second term tends to zero. From their experimental studies on the growth rate of $In_xGa_{1-x}As$ ($x \sim 0.3$) layers on GaAs substrates from In + GaAs solutions, Bolkhovityanov and Zembatov [126] found that the growth rate on (111)A substrates was higher than that on (111)B, and that with values of $D = 4 \times 10^{-4}\ cm^2\ s^{-1}$ and $B = 2.8$ and $1.05 \times 10^{25}\ cm^4\ s^{-1}$ for (111)A and (111)B respectively, the experimental results of $d(t)$ against t could be fitted using equation (3.27). Their explanation for the faster growth on (111)A was that the rate-limiting process was the diffusion of Ga atoms on the crystal surface and that this was likely to be an easier process on the metal-rich (111)A surface than on the As-rich (111)B face.

In considering the relative growth rates of InP, (GaIn)As and GaInAsP on InP substrates of {100}, {111}A and {111}B orientation, Chand *et al* [127] have used the model of stable nucleus formation originally proposed by Sangster [10] and mentioned earlier in Chapter 1. By considering the stable complexes likely to be formed in the solutions Chand *et al* were able to explain their results, and in particular the faster growth on {100} planes than on {111} [127].

More experimental data on the orientation dependence of growth rate as a function of growth temperature and solution supercooling is required before a more satisfactory and complete theoretical explanation can be developed.

3.2.5 Summary
The preceding sections have shown that for a wide variety of growth techniques and conditions, the 'diffusion-limited' model using the 'semi-infinite' solution approximation provides a good description of the LPE growth process, and in particular the time dependence of layer thickness. The theory has also been adapted to explain the formation of the characteristic LPE ripple pattern [24] and the phenomenon of preferential edge growth [122].

3.3 Phase Diagram Information

3.3.1 Introduction

The existence of reasonably accurate liquid–solid phase data for the III–V alloy systems has played an important part in the development of LPE growth of III–V compounds, particularly of the ternary and quaternary systems. The basis of LPE is the control of this liquid–solid phase equilibrium. In the case of the III–V systems, the solid phase is the 1:1 binary compound, e.g. GaAs or an alloy of two or more compounds, while the liquid phase is normally a group III (metal)-rich solution. The important phase information required is:

(i) Liquidus data—the liquidus temperature for a particular liquid-phase composition being the temperature at which, on heating, the two-phase (solid + liquid) mixture becomes a single phase (liquid). LPE growth normally occurs at or slightly below this liquidus temperature.

(ii) Solidus data—this provides information on the composition of the solid phase in equilibrium with the liquid phase at the liquidus temperature.

The following sections discuss the way in which the phase data of relevance to LPE is obtained experimentally, how the data is presented in a form which is directly applicable to LPE growth experiments and finally how theoretical phase-diagram calculations have proved useful in predicting liquidus and solidus data in areas of the phase diagram not explored experimentally.

3.3.2 Experimental phase data

Much experimental liquid–solid phase equilibrium data on III–V alloy systems has been accumulated, and the review of Panish and Ilegems [128] gives an excellent summary of the results up to 1972. Work since then has concentrated on providing more accurate data in certain limited regions of composition and temperature of relevance for specific applications.

Experimental liquidus data has been obtained by three techniques:

(i) 'direct-observation', in which the dissolution of crystals is observed in a semi-transparent furnace, e.g. for the In–P [129] and Ga–In–P [130] systems;

(ii) the measurement of the weight loss from a saturation source placed in contact with a solution of known initial composition, e.g. for the In–Ga–As–P [131,132], In–Ga–As and In–P [132], Ga–Al–Sb [133] and Ga–Al–As [134] systems;

   (iii) differential thermal analyis, e.g. for the Ga–In–As–Sb [135] and Ga–Al–As [136] systems.

Solidus data is normally obtained by measuring the composition of LPE-grown layers by (i) electron-probe micro-analysis, (ii) x-ray diffraction to measure lattice parameter or (iii) luminescence, absorption or reflectivity spectra to measure the bandgap of the material and hence calculate the alloy composition.

Experimental data obtained by these techniques can be extremely accurate and reliable. For example, figure 3.7 is reproduced from the work of Kuphal [132] who compared all the published data for the lattice-matched growth of the $In_{0.53}Ga_{0.47}As$ on (100) InP from seven sources [93,124,125,132,137,138,139]. The agreement is seen to be remarkable and this data can be used with great confidence by new workers. However, it is difficult to obtain such accurate data over the whole of an area of a ternary or quaternary phase diagram.

3.3.3 Presentation of phase data

This section deals with the ways in which phase data is presented in a form which is useful for liquid-phase epitaxy experiments.

From the phase rule introduced in §1.2, $f = c - p + 2$. If pressure P is taken to be a constant, which is a fair assumption for most III–V systems, and since $p = 2$, then the number of degrees of freedom f can be calculated as follows:

binary system $c = 2$ $f = 1$ e.g. Ga–As

ternary system $c = 3$ $f = 2$ e.g. Ga–In–As

quaternary system $c = 4$ $f = 3$ e.g. Ga–In–As–P

For binary systems ($f = 1$), the T–x data can be presented in a two-dimensional figure such as figure 3.8 where data is shown for several binaries. These diagrams are well established for most of the III–V binary systems involving Al, Ga, In and P, As, Sb.

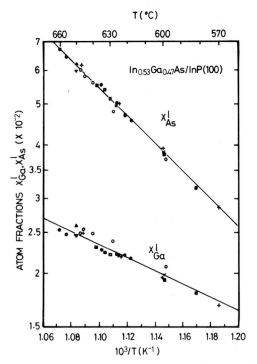

Figure 3.7 Liquid-phase atom fractions X_{Ga}^l and X_{As}^l for lattice-matched growth of $In_{0.53}Ga_{0.47}As$ on InP (100) as a function of temperature from seven sources: (●) Thomson-CSF [124]; (○) Bell Labs [124]; (+) [137]; (△) [138]; (▽) [125]; (□) [93]; (X) [139]; (■) [132] (after Kuphal [132]).

In a ternary system ($f = 2$), e.g. Ga–In–As, the liquid-phase composition is described by three atom fractions, X_{Ga}^l, X_{In}^l and X_{As}^l, the sum of which is unity, while the solid phase is described by a single variable, e.g. x in $Ga_xIn_{1-x}As$. If any two of the four variables (two of the X^ls, x and T) are fixed, the system is defined. For example, if the growth temperature T and solid composition x are chosen, then the liquid phase is uniquely determined. The liquidus in this case is a surface (figure 3.9(a)), and for convenience the data is shown on a Gibb's triangle as liquidus isotherms, such as those shown in figure 3.9(b) for the Ga–In–As system. These are the intersec-

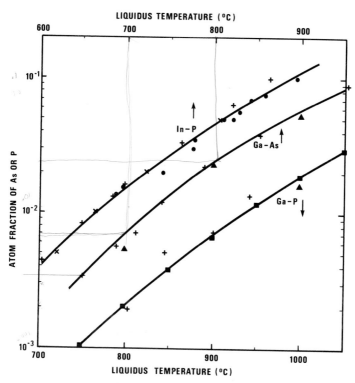

Figure 3.8 Liquidus data for the In–P, Ga–As and Ga–P systems from several sources: (+) Hall [3]; (X) Astles [130]; (●) Panish and Arthur [129]; (▲) Rubenstein [379]; (■) Ilegems *et al* [380].

tions of horizontal temperature planes with the liquidus surface, projected on to the basal plane, as illustrated in figure 3.9(a). The vertical plane between GaAs and InAs is called the 'pseudo-binary' plane (see figure 3.9(a)) representing the continuous solid solution $(GaAs)_x(InAs)_{1-x}$. For LPE applications, solidus data is presented in three ways either as (i) 'isoconcentration' lines on the Gibb's triangle which are the loci of all liquidus compositions in equilibrium with a particular value of alloy composition, x; (ii) more usefully, using rectangular coordinates (figure 3.9(c)) where x is plotted against one of the liquid-phase atom fractions (X^1 in this case) for various temperatures (700 and 800 °C in this case); or (iii) by the use of

(*a*)

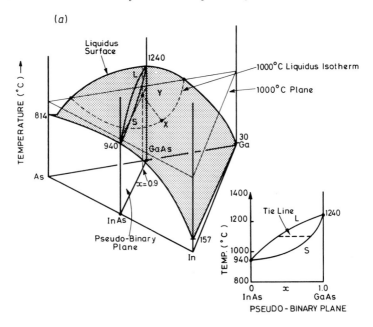

(*b*)

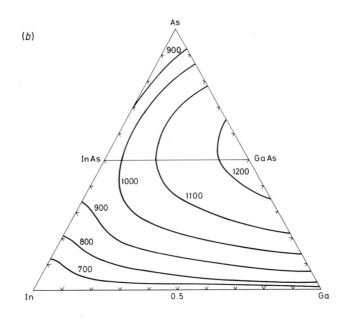

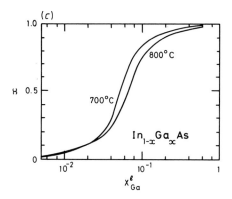

Figure 3.9 (*a*) Liquidus surface for the In–Ga–As system (schematic); (*b*) liquidus isotherms for the In–Ga–As system (700–1200 °C); (*c*) x versus X_{Ga}^l for $In_{1-x}Ga_xAs$ (700, 800 °C) (after Panish and Ilegems [128]).

tie lines which join a liquid composition point with the solid composition x in equilibrium with it, as shown by line XY in figure 3.9(*a*).

For quaternary systems ($f = 3$), e.g. Ga–In–As–P, the liquid phase is characterised by four atomic fractions, e.g. X_{Ga}, X_{In}, X_{As} and X_P whose sum is unity, while the solid phase, e.g. $Ga_xIn_{1-x}As_yP_{1-y}$ is described by the two parameters x and y. In this case, a liquidus isotherm is a surface in a tetrahedral composition figure such as that shown schematically for Ga–In–As–P in figure 3.10. The plane ABCD which passes through the

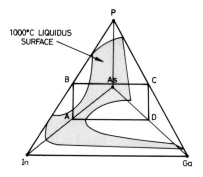

Figure 3.10 Schematic of liquidus isotherm surface for the Ga–In–As–P system at 1000 °C.

InAs, InP, GaP, GaAs points is called the pseudo-quaternary section. Quaternary phase data, experimental and calculated, have concentrated (i) on lattice-matched growth conditions, i.e. where the solid quaternary composition (x, y) can be grown with zero mismatch on to a suitable binary substrate (figure 3.11 shows the lattice-match lines for Ga–In–As–P on to InP and GaAs) substrates and (ii) on the temperature range of interest for LPE, e.g. 500–650 °C for GaInAsP and 500–650 °C for GaInAsSb.

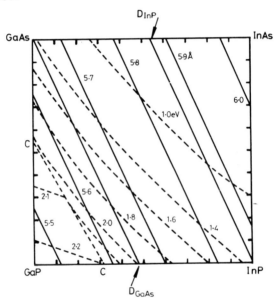

Figure 3.11 Lines of lattice parameter (——) and band-gap (– – –) for the GaInAsP quaternary system showing the lattice-match lines to GaAs and InP substrates.

A convenient way of presenting the liquidus data for lattice-matched growth is on a diagram such as figure 3.12 where the ordinates are the various liquidus atom fractions, X_1^l, and the ordinate is the solid composition y. For lattice-matched growth this value of y determines the other solid parameter x. If the growth temperature, T, is then chosen, the liquidus composition is uniquely determined.

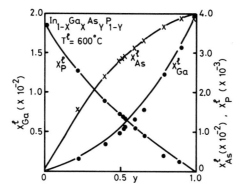

Figure 3.12 Liquid-phase atom fractions X^l_{Ga}, X^l_{As} and X^l_P for lattice-matched growth of $In_{1-x}Ga_xAs_yP_{1-y}$ onto InP at 600 °C. Curves are calculated from polynomial functions, and data points are experimental values (after Kuphal [132]).

Because of the difficulty in obtaining experimental phase-diagram information it has been found advantageous to perform theoretical calculations, particularly in the early stages of research on a new material where the exact growth conditions (composition, temperature) are not yet fixed. The next section describes these theoretical calculations.

3.3.4 Calculations of III–V phase equilibria

Because the calculations of ternary III–V phase equilibrium have been well reviewed by Panish and Ilegems [128], this section will not attempt to provide a detailed account of the subject. Instead, we shall merely discuss the results of calculations for a few III–V systems and show how the simple models can be extended (i) to quaternary phase diagram calculations and (ii) to take account of association in the liquid phase. The basic approach is to derive equations for the chemical potentials of the species in the liquid and solid phases (μ^l, μ^s). At equilibrium the chemical potentials of a species in the solid and liquid phases can be equated, e.g.

$$\mu^l_A = \mu^s_A.$$

Regular solution model

Most calculations of III–V phase diagrams have been based on a regular solution model of the liquid and solid phases.

(i) *Binary systems*. For a binary solution A + B (e.g. Ga–As) the free energy of mixing $\Delta G_M^{(l)}$ is given by

$$\Delta G_M^{(l)} = RT[X_A^l \ln X_A^l + X_B^l \ln X_B^l] + \alpha_{AB} X_A^l X_B^l \quad (3.28)$$

where X_A^l, X_B^l are the atom fractions of A, B in the liquid phase ($X_A + X_B = 1$), and α_{AB} is the interaction parameter between A and B.

The final term in (3.28) is called the excess free energy of mixing (ΔG_M^{xs}). For most III–V metal-rich solutions, α has been found to have the form

$$\alpha = a - bT \quad (3.29)$$

and such solutions are called 'quasi-regular'.

The chemical potentials of the species in the liquid phase, e.g. μ_A^l are given by

$$\mu_A^l = RT \ln \gamma_A^l X_A^l$$

(and similarly for B) where γ_A^l is the activity coefficient of A in the liquid given by

$$RT \ln \gamma_A^l = \alpha_{AB}^l (X_B^l)^2$$
$$RT \ln \gamma_B^l = \alpha_{AB}^l (1 - X_B^l)^2. \quad (3.30)$$

Vieland [140] has derived the following approximate expression for α for a binary system using a regular solution treatment

$$\alpha_{AB} = \frac{-RT}{2(0.5 - X_B^l)^2} \left[\ln 4 X_A^l X_B^l + \frac{\Delta S_{AB}^f}{R} \left(\frac{T_{AB}^f - T}{T} \right) \right] \quad (3.31)$$

where ΔS_{AB}^f is the entropy of fusion of compound AB and T_{AB}^f is the melting point of AB.

Sets of values of the parameters such as ΔS^f, T^f and α for many III–V systems are given in the review by Panish and Ilegems [128].

(ii) *Ternary systems*. We now consider a ternary system A–B–C (e.g. Ga–In–As) where the solid phase is the pseudo-binary alloy $A_x B_{1-x} C$.

The equations presented previously for binary solutions can be extended to ternary solutions.

The chemical potentials in the liquid and solid phases can be written as follows:

$$\text{liquid:} \quad \mu^l_A = \mu^{lo}_A + RT \ln \gamma^l_A X^l_A \qquad \text{etc for B, C}$$

$$\text{solid:} \quad \mu^s_{AC} = \mu^{so}_{AC} + RT \ln \gamma^s_{AC} x \qquad (3.32)$$

$$\mu^s_{BC} = \mu^{so}_{BC} + RT \ln \gamma^s_{BC}(1 - x)$$

where μ^{lo}_A is the chemical potential of pure liquid A, μ^{so}_{AC} is the chemical potential of pure solid AC and x is the solid alloy composition.

The activity coefficients, γ, are given by

$$RT \ln \gamma^l_A = \alpha_{AB}(X^l_B)^2 + \alpha_{AC}(X^l_C)^2$$

$$+ (\alpha_{AB} - \alpha_{BC} + \alpha_{AC})X^l_B X^l_C \quad \binom{\text{similarly}}{\text{for B, C}} \quad (3.33)$$

$$RT \ln \gamma^s_{AC} = \alpha^s_{AC\text{-}BC}(1 - x^2) \qquad \text{(similarly for B, C).} \,(3.34)$$

In published work the activity coefficients of the solid phase have been either (i) taken as unity, i.e. solid solution AC–BC is assumed ideal, (ii) calculated from (3.34) with a temperature independent $\alpha_{AC\text{-}BC}$ or (iii) fitted to a polynomial of the form

$$RT \ln \gamma^s_{AC} = Ax^2 + Bx + C.$$

Which of these is most appropriate depends on the alloy system.

The equilibrium conditions are as follows:

$$\mu^s_{AC}(T) = \mu^l_{AC}(T) = \mu^l_A(T) + \mu^l_C(T)$$

$$\mu^s_{BC}(T) = \mu^l_{BC}(T) = \mu^l_B(T) + \mu^l_C(T). \qquad (3.35)$$

By making use of Vieland's expression for the chemical potentials of the pure compound μ^{so}_{AC}

$$\mu^{so}_{AC}(T) = \mu^{sl}_A(T) + \mu^{sl}_C(T) - \Delta S^f_{AC}(T^f_{AC} - T) \qquad (3.36)$$

where the μ^{sl} terms represent the chemical potentials in the stoichiometric liquid, and substituting (3.32) and (3.36) into (3.35), the following equations are obtained:

$$\gamma^s_{AC} x = 4 X^l_A X^l_C \frac{\gamma^l_A \gamma^l_C}{\gamma^{sl}_A \gamma^{sl}_C} \exp\left(\frac{\Delta S^f_{AC}(T^f_{AC} - T)}{RT}\right)$$

$$\gamma^s_{BC}(1 - x) = 4X^l_B X^l_C \frac{\gamma^l_B \gamma^l_C}{\gamma^{sl}_B \gamma^{sl}_C} \exp\left(\frac{\Delta S^f_{BC}(T^f_{BC} - T)}{RT}\right). \quad (3.37)$$

These equations can be solved numerically for x and T for various input values of X^l_A, X^l_B, X^l_C. The main problem in performing these calculations at present is that the values of some of the critical parameters (interaction parameters, entropies of fusion) are not well known. Generally, authors have fitted the calculated results to established experimental data by adjusting the parameters in the calculation. The region of the phase diagram around the 'fit point' can then be calculated with fair accuracy. Figure 3.13 shows a comparison of solidus data obtained experimentally by growth of (GaIn)As on 111(B) InP with calculated values. The agreement is good, although the authors admit that tailoring of some of the input parameters was necessary to get a good fit [92].

(iii) *Quaternary systems.* The recent upsurge in activity on the

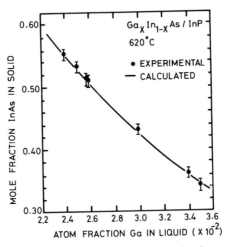

Figure 3.13 Comparison of calculated (——) and experimental (●) results for the variation of $(1 - x)$ with X^l_{Ga} for growth of $Ga_x In_{1-x}As$ onto InP at 620 °C (after Pearsall and Hopson [92]).

LPE growth of quaternary III–V alloys, and the great amount of work involved in carrying out experimental determinations of even quite small regions of the phase diagram, have led to an interest in applying the theoretical treatment just described for ternaries, to the calculation of quaternary phase diagrams. The model for the ternary liquid phase outlined in the previous section can be easily extended to a four-component solution. The activity coefficients γ_i^l are given by

$$RT \ln \gamma_i = \sum_{\substack{j=1 \\ j \neq i}}^{4} \alpha_{ij}(X_j^l)^2 + \sum_{\substack{j=1 \\ j \neq i}}^{4} \sum_{\substack{k=1 \\ k=ij}}^{4} X_j^l X_k^l (\alpha_{ij} + \alpha_{ik} - \alpha_{jk}). \quad (3.38)$$

The main problem in quaternary phase diagram calculations has been in calculating the activity coefficients of the solid phase, γ_{ij}^s, in the case of III/III′/V/V′ quaternaries (e.g. GaInAsP, GaAlAsSb). The problem arises from the fact that for a given solid-phase composition, e.g. given by x and y in $In_{1-x}Ga_xAs_{1-y}P_y$, the free energy of mixing ΔG_M is not a straightforward function of x and y, since the solid-phase composition can be derived in an infinite number of ways from the four binaries (GaAs, GaP, InAs, InP). While some workers have assumed that the quaternary solid is ideal, i.e. $\gamma_{ij}^s = 0$, this has been found to be a poor approximation. Stringfellow [141] has used the so-called 'delta lattice parameter' model to calculate the free energy of mixing. Using this approach, Stringfellow has calculated liquidus and solidus points for the GaInAsP system [141] and found generally poor agreement with experiment. An ideal solid solution model was used by Antypas and Edgecumbe [142] and while relatively poor agreement with liquidus experimental data was obtained, reasonable fits to the distribution coefficients of P, Ga and As as a function of temperature were obtained.

The main uncertainties in these calculations are the values of the interaction parameters, α^l, which are obtained by fitting to experimental binary data. However, this data is often in a temperature range higher than that of interest in the quaternary calculations and the linear relationship of equation (3.29) may not hold over a wide temperature range. Generally, the fit of calculated quaternary phase data to experimental values has been poor in comparison with ternaries. Recent low-temperature liquidus measurements in the binary systems Ga–P, In–P, Ga–As and In–As [143] should enable better values of α^l to be used in the calculation of the Ga–In–As–P system. Another approach used by some authors [132] is to fit the experimental data to a set of polynomials. These can then be used to interpolate between or extrapolate from the range of experimental measurements. An example of this is shown in figure 3.12 where the experimental values of X^l_{Ga}, X^l_{As} and X^l_P for lattice-matched growth of GaInAsP onto (100) InP at 600 °C are compared with those calculated using the empirical polynomial functions [132].

3.3.5 Association effects in the liquid phase

One of the basic assumptions of the calculations outlined above is that the liquid phase can be considered as consisting of discrete atomic species. There is experimental evidence from measurements of surface tension [144,145], viscosity and electrical conductivity [146] on the group III antimonides (AlSb, GaSb and InSb) for significant association at the melting point. Recently Szapiro [147] has extended the regular-associated solution (RAS) concept of Jordan [148] to ternary solutions, in particular to the Ga–In–Sb system. The model considers a ternary solution containing atomic species A, B, C and molecular species AC, BC such that

$$AC \rightleftarrows A + C \qquad K_1 = a_A a_C / a_{AC}$$
$$BC \rightleftarrows B + C \qquad K_2 = a_B a_C / a_{BC}. \qquad (3.39)$$

Szapiro [147] effectively calculated K_1 and K_2 by fitting to experimental points at various temperatures. However, the extremely good fits to experimental liquidus and solidus isotherms over a wide range of temperatures suggest that this model is an improvement over previous 'simple' solution treatments of the Ga–In–Sb system.

Linnebach [149] has used a similar treatment for the Al–Ga–Sb ternary system and was able to obtain a good fit to both ternary data and also the binary systems on both the metal-rich and Sb-rich sides using only one set of parameters, instead of the two separate sets required in the simple-solution model. It would be interesting to try to tie in the RAS model with more detailed experimental studies of the degree of association in III–V solutions.

3.3.6 *Substrate instability*

In the heteroepitaxial growth of III–V materials, the situation can arise whereby the liquid phase to grow a particular layer comes into contact with a substrate or previously grown layer with which it is not strictly in equilibrium. The way in which this non-equilibrium situation approaches equilibrium varies from one material system to another. In the worst case, the underlying material can be severely eroded by the solution, for example in the case of (Ga, In)As on InAs [17]. Bolkhovityanov *et al* [150] have examined the instability behaviour of many III–V alloy systems and formulated a model and criteria which could predict whether a given liquid–solid combination would be stable or not. The postulated initial process which occurs when the liquid comes into contact with the solid is that a thin film is formed either by solid state diffusion or by a process of 'etch-back and regrowth' or fluctuation exchange. In lattice-matched growth situations such as Ga–Al–X solution in contact with GaX substrates (X = P, As, Sb) the formation of a diffusive 'skin' is favoured. Without the formation of this 'skin', dissolution of the substrate (GaX) would be expected, the tendency increasing with X going from P to As to Sb. This is due to the fact that the dissolution of small, equimolecular amounts of Ga and X causes a decrease in the liquidus temperature and hence an increased tendency to further dissolution. In the case of X = Sb, the erosion process dominates the diffusive skin formation and substrate erosion is found.

In the case where the equilibrium solid is lattice mismatched to the substrate another factor needs to be considered. This is the strain energy in the diffused skin, which introduces a new energy term in the expression for the free energy of the solid phase. This strained solid is no longer in equilibrium with the liquid phase, and the latter needs to be supercooled by ΔT_E (the stress-induced supercooling) in order to re-establish equilibrium. If the liquid is maintained at the original equilibrium temperature then etch-back of the substrate occurs. If this substrate dissolution changes the liquid composition in such a way that the new equilibrium solid has a lower mismatch to the substrate, then the driving force for dissolution is reduced and vice versa. For example, in the growth of InGaAsP on an InP substrate, if the initial $\Delta a/a$ is positive, then dissolution of the InP substrate leads to an increased In, P content in the solution, and since $k_P \gg k_{In}$, then the new equilibrium solid will have a reduced lattice parameter, the values of $\Delta a/a$ will decrease and any tendency to substrate dissolution will be decreased. For an initial negative $\Delta a/a$, the substrate dissolution would be expected to be unstable, since each stage of dissolution leads to an equilibrium solid of higher negative mismatch.

Using the three criteria of (i) the sign of the initial mismatch $\Delta a/a$, (ii) whether or not dissolution leads to an increase or decrease in $\Delta a/a$ for the new equilibrium solid, and (iii) the effect of dissolution on the liquidus temperature of the new solution composition, Bolkhovityanov has been able to explain the dissolution behaviour of all the III–V binary, ternary and quaternary systems for which there are experimental observations.

3.4 Impurity Incorporation

The understanding of the mechanisms of impurity incorporation into epitaxial films is obviously an important prerequisite for the growth of device material where there is likely to be stringent specifications on doping levels and uniformity in the device structure. The subject is highly complex, and in this section only an outline will be given of the various theories and the circumstances in which they are applicable.

Most of the published work on doping of LPE layers of III–Vs is purely empirical in nature, and in only a few cases has a detailed study been made of the effect of the various growth parameters, i.e. growth temperature (T_G), growth rate (v), dopant concentration in the liquid (X_1^l) and crystallographic orientation of the growing layer.

When trying to draw conclusions on impurity segregation in semiconductors from the results of electrical carrier concentration measurements, certain assumptions are generally made:

(i) The impurity atoms are neutral or singly ionised.

(ii) The impurity atoms are fully ionised at room temperature.

(iii) The concentration of compensating complex centres involving the impurity is negligible.

(iv) The dopant concentration is uniform through the layer.

(v) Precipitation has not significantly occurred on cooling the crystal from its growth temperature to room temperature.

In some work, these assumptions have been investigated, for example by depth profiling (to check on (iv)), or by analysis of the Hall mobility data to obtain N_d, N_a values (to check on (iii)). Alternatively, several authors have used chemical analysis of the solid to measure directly the impurity concentration. For example, Casey *et al* [43] measured the radioactivity of GaAs layers doped with the isotope ^{129m}Te to obtain the total Te concentration C_{Te} and found that above $C_{Te} \sim 5 \times 10^{18}$ cm^{-3}, the carrier concentration n was significantly less than C_{Te}. Unfortunately, this technique cannot be applied to many dopants of interest because of the lack of isotopes with suitable half-lives.

More recently, secondary-ion mass spectroscopy (SIMS) has begun to be used to provide accurate chemical analysis of impurities in epitaxial layers.

Equilibrium theory

The incorporation of impurities is usually described in terms of the so-called distribution or segregation coefficient, k, defined as

$$k = C_S/C_L$$

where C_S and C_L are the concentrations of the impurity in the solid and liquid phases, respectively.

As shown in figure 3.14, this definition can be applied only when the growth rate v is extremely slow compared to the rate of diffusion of the impurity in the liquid phase, so that the impurity distribution in both solid and liquid is flat (figure 3.14(*a*)). When this condition is not met, the situation shown in figure 3.14(*b*) occurs and two values of k can be defined:

$$k_o = C_S/C_L(0) \quad \text{at the surface}$$
$$k_{eff} = C_S/C_L(\infty) \quad \text{in the bulk.}$$

Generally the growth rates in LPE are sufficiently low for it to be believed that $k_o \sim k_{eff}$, and that equilibrium theory can be applied.

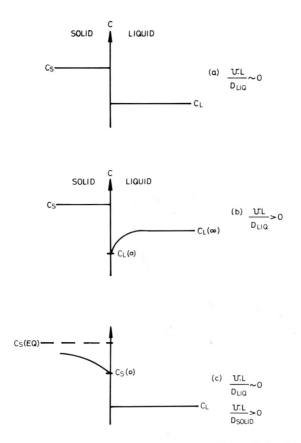

Figure 3.14 Impurity profiles in the solid and liquid phases for (*a*) very slow growth, (*b*) faster growth and (*c*) very slow diffusion in the solid phase. The growth rate is v, the impurity diffusion coefficient in the solution is D_{LIQ} and the solution thickness is L.

The equilibrium theory has been widely used to describe impurity incorporation into LPE layers. As applied to Sn in GaAs, for example,

$$Sn(l) + V_{Ga} \rightleftarrows Sn_{Ga}^+ + e^- \qquad (3.40)$$

where Sn_{Ga} represents a tin atom on a Ga site in the lattice, V_{Ga} represents a gallium vacancy at the surface of the growing layer and $Sn(l)$ is a tin atom in the liquid phase. The equilibrium coefficient K_e for (3.40) is given by

$$K_e = \frac{[Sn_{Ga}^+]n}{[Sn(l)][V_{Ga}]} \qquad (3.41)$$

where [ ] represents concentration.
The equation $e^- + h^+ \rightleftarrows 0$ also applies, with

$$K_i = n_i^2 = np$$

where n and p are the electron and hole concentrations in the solid and n_i is the intrinsic carrier concentration at the growth temperature. For low $[Sn_{Ga}^+]$, $n = n_i \gg [Sn_{Ga}^+]$:

$$[Sn_{Ga}^+] = \frac{K_e[V_{Ga}][Sn(l)]}{n_i} = K_{int}[Sn(l)]. \qquad (3.42)$$

For high $[Sn_{Ga}^+]$, where $[Sn_{Ga}^+] = n \gg n_i$:

$$[Sn_{Ga}^+] = [K_e[V_{Ga}][Sn(l)]]^{1/2} = K_{ext}[Sn(l)]^{1/2}. \qquad (3.43)$$

The transition from a linear relationship (3.42) to a square-root one (3.43) has been observed, e.g. Te in GaP [53] at 1040 °C. However, the transition occurs at a doping level of $\sim 6 \times 10^{17}$ cm^{-3}, which is significantly higher than n_i ($\sim 1.5 \times 10^{17}$ cm^{-3} at 1040 °C). For other systems, e.g. Te in GaAs, no switch to a square-root relationship has been seen up to very high carrier concentrations [43].

Several other pieces of work have also cast doubt on the applicability of the equilibrium model. König has shown that the incorporation of Sn in heavily doped GaAs increases with growth rate v in the range 0.2 to 6 μm min^{-1} [151]. Kang and Greene [40] found that the incorporation of Sn in GaAs was orientation dependent ($k_{(111)B}^{Sn} \sim 4k_{(111)A}^{Sn}$), while Keller and Muench [33] found for Zn in GaAs that ($k_{(111)A}^{Zn} \sim 2k_{(111)B}^{Zn}$).

To try to explain these results, several theories have been proposed which can be grouped as surface incorporation theories.

Surface-controlled incorporation

The basis of this incorporation model is that if the diffusion of the impurity atom in the solid phase is slow while the liquid phase is close to equilibrium, the surface of the solid phase is not at equilibrium with the bulk of the solid (figure 3.14(c)). In this situation, it has been proposed that the position of the Fermi level at the surface controls the incorporation of the impurity. Several workers have considered the semiconductor–solution interface as a Schottky barrier [125], so that the electron concentration at the surface n_S is given by

$$n_S \sim N_C \exp\left(-\Phi_B/kT\right) \qquad (3.44)$$

where N_C is the effective density of states and Φ_B is the barrier height as shown in figure 3.15. Thus n_S is a function of temperature only. If equation (3.44) is substituted into equation (3.41) for K_e then

$$K_e = \frac{[Sn_{Ga}^+]n_S}{[Sn(l)][V_{Ga}]} \qquad (3.45)$$

thus

$$[Sn_{Ga}^+] = K_{SB}[Sn(l)]. \qquad (3.46)$$

This model thus predicts a linear relationship between the solid- and liquid-phase dopant concentrations, irrespective of the doping level. To explain why some dopants exhibit a square-root relationship at high doping levels as expected from the bulk equilibrium model, the surface-controlled model postulates that the diffusion coefficient of the dopant in the solid phase (D_S) is the important parameter, and that if the diffusion coefficient increases at high doping level, the surface-controlled model no longer holds, and the bulk equilibrium model again becomes valid.

The orientation effects are explained on the basis of different values of Φ_B on different crystal faces, which has been found for Au Schottky barriers on GaAs, for example.

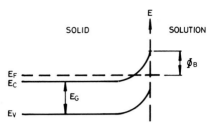

Figure 3.15 Schottky barrier model for impurity incorporation in LPE growth.

In the absence of firm experimental evidence on the nature of the metal(solution)/semiconductor interface at high temperatures, this model is speculative, but it is clear that under a wide range of growth conditions, true equilibrium between the bulk solid and liquid phases does not occur.

Chapter 4

Materials Parameters and Structural Characteristics of LPE III–V Materials

4.1 Introduction

In this chapter, several properties and characteristics of LPE-grown III–V materials that are of relevance to device applications will be discussed under the following headings:

Bandgaps	(4.2)
Electrical properties	(4.3)
Optical properties	(4.4)
Crystal perfection	(4.5)
Layer thickness control	(4.6)
Abrupt junctions	(4.7)
Selective area and structured substrate growth	(4.8)
Composition control	(4.9)

These topics are all of relevance to the device engineer as they affect such device characteristics as spectral response, operating frequency, bandwidth, quantum efficiency etc. Many of these materials parameters are routinely measured in many laboratories to assist in materials and device research or to screen material for device fabrication. The wide range of III–V materials and the differing results from various laboratories make a comprehensive treatment impossible. Instead, general comments will be made and typical examples used for illustration. Comparisons between the properties of material grown by LPE and that grown by other epitaxial techniques are made wherever possible.

4.2 Bandgaps

The band structure of a semiconductor determines many key parameters, such as the optical bandgap, the effective mass of holes and electrons, carrier mobilities, velocity field characteristics etc. Since the main emphasis in LPE materials is on optoelectronic devices, we shall consider the bandgap as an optical parameter.

If we consider the binary III–V compounds formed from the group III elements Al, Ga, In and the group V elements N, P, As, Sb we have a total of 12 materials whose minimum room-temperature energy gaps are shown in table 4.1. It should be pointed out that, apart from AlN, the Al compounds are unstable in moist air and, compared to the Ga and In compounds, have been little investigated in binary form, although they have been commonly grown as components of ternary and quaternary alloys.

Table 4.1 Room-temperature bandgaps of binary III–V compounds (eV). (D = direct gap, I = indirect gap, * grown by LPE.)

AlN	AlP	AlAs*	AlSb
6.02(I)	2.45(I)	2.16(I)	1.62(I)
GaN*	GaP*	GaAs*	GaSb*
3.34(D)	2.26(I)	1.44(D)	0.7(D)
InN	InP*	InAs*	InSb*
2.0(D)	1.35(D)	0.36(D)	0.17(D)

It can be seen that the lower bandgaps tend to be direct in character, where the minimum in the conduction band and the maximum in the valence band are at the same position in k-space. The higher-bandgap materials tend to have indirect gaps. In alloy systems where one of the limiting binaries has a direct gap and the other an indirect gap, there will be a 'cross-over' at some intermediate composition, as shown in figure 4.1(a) for the $In_{1-x}Al_xP$ system.

Of the ternary III–V alloys, many have been grown over at least part of the range of alloy composition by LPE. The (As, Sb) and (P, Sb) systems have regions of immiscibility which

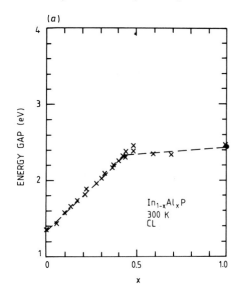

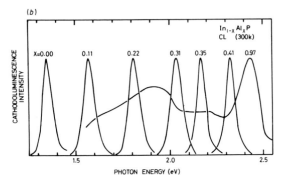

Figure 4.1 Room-temperature cathodoluminescence results for $In_{1-x}Al_xP$ alloys as a function of x, showing
(*a*) band structure with direct–indirect crossover at
$x = 0.44$, 2.33 eV and (*b*) corresponding CL spectra at
300 K (both after Onton and Chicotka [381]).

restrict the range of compositions which can be grown by LPE.
Table 4.2 summarises the range of alloy composition and the
corresponding bandgap range that have been grown in these

systems. In most of these alloy systems, the major limitation on the range of alloys (bandgaps) obtainable, other than miscibility gaps, is the problem of lattice matching to an available binary substrate. In spite of these limitations, it can be seen from table 4.2 that a wide range of energy gaps is available.

Table 4.2 Available bandgaps in LPE-grown ternary III–V alloys.

Alloy system	Bandgap range (300 K) grown by LPE (eV)	Composition range (x, y)	Substrate	Reference
$InAs_yP_{1-y}$	0.5 –1.34	$y = 0$–0.74	InP	[153]
$GaAs_{1-y}Sb_y$	1.25–1.44	$y = 0$–0.11	GaAs	[154]
	1.16–1.44	$y = 0$–0.17	GaAs	[94]
$InAs_{1-y}Sb_y$	0.24–0.36	$y = 0$–0.2	InAs	[155]
	0.27–0.29	$y = 0.10$–0.14	GaSb	[156]
$Ga_{1-x}Al_xP$	2.25–2.4	$x = 0$–0.86	GaP	[157]
$Ga_xIn_{1-x}P$	2.25	$x = 0.8$–1.0	GaP	[158]
	1.85	$x = 0.52$	GaAs	[159]
$Ga_{1-x}Al_xAs$	1.44 to ~2.1	$x = 0$ to ~0.9	GaAs	[160]
$In_{1-x}Al_xAs$	1.55	$x = 0.48$	InP	[161]
$In_{1-x}Ga_xAs$	1.1 –1.44	$x = 0.8$–1.0	GaAs ⎱	[162]
	0.36–0.58	$x = 0$–0.3	InAs ⎰	
	0.72–0.81	$x = 0.45$–0.54	InP	[163]
$Ga_{1-x}Al_xSb$	0.68–1.5	$x = 0$–0.8	GaSb	[133]
$In_{1-x}Ga_xSb$	0.17–0.4	$x = 0$–0.59	InSb	[164]
	0.17–0.62	$x = 0$–0.9	InSb ⎱	[165]
	0.43–0.68	$x = 0.65$–1.0	GaAs ⎰	

When quaternary III–V alloys are considered, independent choice of bandgap and lattice parameter is available. This can be seen in figure 4.2 for the system $Ga_{1-x}In_xAs_{1-y}P_y$, where the full lines represent constant lattice parameter, and the broken lines are contours of constant bandgap. The horizontal and vertical axes represent (x, y), the alloy composition parameters. The lines labelled D_{InP} and D_{GaAs} represent the locus of alloy compositions (x, y) which are lattice matched to InP and GaAs respectively.

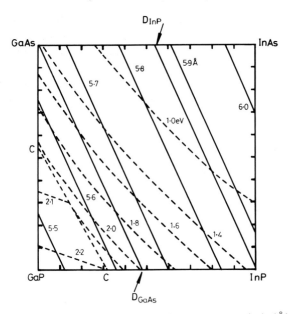

Figure 4.2 Lines of constant lattice parameter (—) (Å) and energy gap (- - -) (eV) for the InGaAsP quaternary system showing the lattice-match lines for GaAs and InP substrates and the transition (C–C) between direct and indirect energy gaps.

There are 15 quaternary III–V alloy systems involving the group III elements Al, Ga, In and the group V elements P, As, Sb. These can be subdivided into three categories:

(i)   *III/III′/III″/V alloys*
    (AlGaIn)P; (AlGaIn)As; (AlGaIn)Sb
(ii) *III/III′/V/V′ alloys*
    (AlGa)(PAs); (AlGa)(PSb); (AlGa)(AsSb);
    (GaIn)(PAs); (GaIn)(PSb); (GaIn)(AsSb);
    (AlIn)(PAs); (AlIn)(PSb); (AlIn)(AsSb)
(iii) *III/V/V′/V″ alloys*
    Al(PAsSb); Ga(PAsSb); In(PAsSb)

Of these, several have not been investigated in any detail either because they have no lattice-matched compositions with interesting properties or because they are difficult to grow. For example, the PSb-containing alloys are likely to have widespread regions of immiscibility. Figures 4.2–4.6 show the ranges of lattice-matched compositions which can be grown in the quaternary systems GaInAsP (4.2), GaInAsSb (4.3), GaAlAsP (4.4), GaAlAsSb (4.5) and InPAsSb (4.6), and most of these have actually been grown by LPE at least over part of the range.

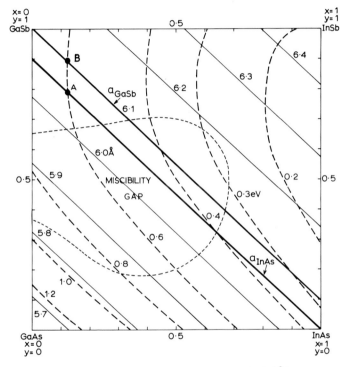

Figure 4.3 Lines of constant lattice parameter (—) (Å) and 300 K energy gap (- - -) (eV) for the $Ga_{1-x}In_xAs_{1-y}Sb_y$ quaternary system showing the lattice-match lines for InAs and GaSb substrates.

Table 4.3 shows the ranges of bandgaps of several quaternary systems grown by LPE near lattice matched to the substrates indicated.

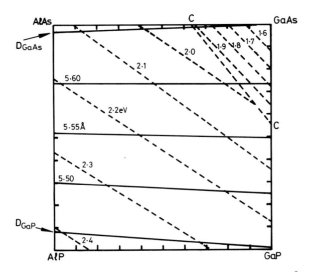

Figure 4.4 Lines of constant lattice parameter (—) (Å) and 300 K energy gap (- - -) (eV) for the GaAlAsP quaternary system, showing the lattice-match lines for GaAs and GaP substrates and the transition (C–C) from direct to indirect energy gaps.

It can thus be seen that by the use of binary and ternary III–V alloys, an extremely wide range of bandgaps can be obtained in the visible and near infrared region. Quaternary alloys provide the possibility of growing lattice-matched layers of a desired bandgap on a binary substrate.

4.3 Electrical Properties

The measurement of electrical properties of epitaxial layers is probably the most common type of assessment, since the properties measured, such as carrier concentration and type, mobility, deep level properties, are crucial to the operation of electronic devices. In this section, typical LPE results will be briefly described.

4.3.1 Background purity
To achieve control over the electrical properties of semiconduc-

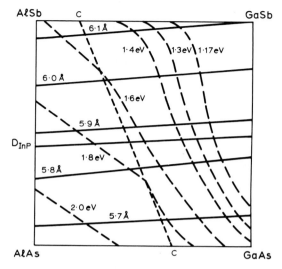

Figure 4.5 Lines of constant lattice parameter (—) (Å) and 300 K bandgap (- - -) (eV) for the AlGaAsSb quaternary system showing the lattice-match line for an InP substrate and the transition (C–C) from direct to indirect energy gaps.

tor materials, it is firstly essential to reduce as far as possible any unintentional impurities. In the case of LPE there are three major sources of impurities in the growth process. These are:

(i) impurities in the solution starting materials;
(ii) impurities emanating from the container material (normally graphite);
(iii) impurities in the ambient gas.

The first of these three sources is the one which has been the subject of most research. In the case of GaAs grown from Ga solutions, the dominant impurities are generally accepted to be C (donor), O (donor) and Si (donor or acceptor) [39,69, 182–197]. Of these, oxygen has been the main target for removal. Baking of the growth solutions in H_2 has been the favoured way of reducing the oxygen concentration, the important factors being the bake temperature, the bake time and the O_2 concentration of the input H_2 gas. Several workers agree

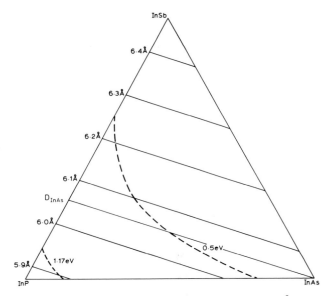

Figure 4.6 Lines of constant lattice parameter (—) (Å) and 300 K energy gap (- - -) (eV) for the InPAsSb quaternary system.

that at a fixed growth temperature, there is a decrease in $(N_D - N_A)$ in undoped LPE GaAs layers as the bake temperature increases [183,184,188] to the extent that a switch from n- to p-type at the highest bake temperatures occurs. Analysis of the 77 K Hall data carried out by Nanishi [183] suggests that N_A stays constant and N_D decreases with increased bake temperature. This has been interpreted by these workers as being a steadily decreasing equilibrium oxygen content in the Ga solution as the bake temperature increases. In fact, more recent work by Chang *et al* [186], who measured the oxygen activity in liquid Ga by an electrochemical technique, suggests that this model is incorrect. With increasing bake temperature, T_B, the equilibrium oxygen activity actually increases although the equilibrium value is attained more rapidly. However, Otsubo *et al* [69] have shown that the distribution coefficient at O, k_O, in GaAs LPE falls from 6.5×10^{-4} at 700 °C to 5.1×10^{-5} at 800 °C. From the figures in the papers of Chang *et al* [186] and

Table 4.3 Bandgap ranges available in lattice-matched III–V quaternary alloys grown by LPE. (D = direct gap, I = indirect gap.)

Material	Range of bandgaps at 300 K (eV)		Lattice match to	Reference
GaAlAsP	2.25–2.4	(I)	GaP	[166]
	1.44–1.9	(D)	GaAs	[167]
	1.9–2.16	(I)		
GaAlAsSb	0.7–1.5	(D)	GaSb	[168]
	1.5–1.65	(I)		
	1.18–1.58	(D)	GaAs	[169]
GaAlInAs	0.7–1.5	(D)	InP	[170]
			GaAs	[171]
GaAlInP	1.91–2.17	(D)	GaAs	[172]
	2.17–2.25	(I)		
GaInAsP	0.7–1.35	(D)	InP	[173]
	1.44–2.0	(D)	GaAs	[174]
	1.874–2.115	(D)	$GaAs_yP_{1-y}$ ($y \sim 0.69$)	[175]
	1.94–2.0	(D)	$GaAs_yP_{1-y}$ ($y \sim 0.61$)	[176]
GaInAsSb	0.36–0.7	(D)	InAs	[177]
	0.53–0.72	(D)	GaSb	[178]
InAsPSb	~0.3–0.5	(D)	InAs	[179]
GaAsPSb	Graded composition		GaAs	[180]
GaAlInSb	~1.6		GaSb	[181]

Otsubo *et al* [69], the product of $k_O \times [O]_{Ga}^{eq}$ decreases with increasing temperature, where $[O]_{Ga}^{eq}$ is the equilibrium concentration of oxygen in liquid Ga. Thus the incorporation of oxygen donors into the LPE layer would decrease with increasing growth temperature T_G, as observed experimentally [183,184,188]. The variations between different authors on the optimum baking time and temperature to minimise $N_D - N_A$ reflects the influence of other factors such as the oxygen concentration in the incoming H_2 gas and the accumulation of impurities from the boat material during the bake period. Other ways of reducing the oxygen content of the growth solution have also been studied. Shealy and Woodall [189] developed a technique to remove water vapour and oxygen from hydrogen or nitrogen gases to below 1 PPM by bubbling the gas through a Ga + In + Al mixture which is molten at room temperature.

The addition of Ti and Zr to Ga solutions as oxygen getters has been investigated by Stevenson *et al* [190]. With the addition of Ti, low 10^{14} cm^{-3} n-type GaAs was obtained for low pre-growth solution baking temperatures and $\sim 10^{15}$ cm^{-3} p-type for higher bake temperatures. Although the levels of Ti and Zr incorporated into these layers were measured in this work, the authors did not reveal the results in the paper. It is thus difficult to assess whether the dominant effect is a gettering action in the solution or the introduction of deep levels associated with Ti, Zr in the GaAs layer. However, Chang *et al* [186] subsequently measured electrochemically the reduction in oxygen activity in Ga solutions when Zr was added and noted a rapid reduction in activity to a level nearly four orders of magnitude lower than that achieved by prolonged H$_2$ baking.

Contamination of LPE GaAs with Si has been found to be a problem only when the growth is carried out in silica boats. The concentration of Si in liquid Ga from this source can be reduced through the reaction

$$2H_2 + SiO_2(s) \rightleftarrows Si(l) + 2H_2O(g) \qquad (4.1)$$

$$\underset{\substack{\text{silica} \\ \text{boat}}}{\uparrow} \qquad \underset{\substack{\text{dissolved} \\ \text{in Ga}}}{\uparrow}$$

by increasing the concentration of water vapour in the H$_2$ gas. This has been done by Hicks and Greene [191] who achieved low 10^{13} cm^{-3} n-type material with $\mu_{77} > 10^5$ cm^2 V^{-1} s^{-1}.

The situation in the case of In-containing materials such as InP, InGaAs and InGaAsP grown from In-rich solutions is slightly different. The major background impurities are Si (donor), S (donor) and Zn (acceptor). Kuphal and Pöcker [192] measured the impurities in the high-purity In by atomic absorption spectroscopy (AAS). The commonly detected impurities were Mg, Al and Si, all of which were reduced by a prolonged bake in dry H$_2$. The distribution coefficient of Si is so high [46, 50] that even concentrations of ~ 0.001 PPMA in the In solution could give rise to the background donor carrier concentration of $1–2 \times 10^{15}$ cm^{-3} commonly found in LPE InP [193–195]. Holmes and Kamath [196] have measured Si and S directly in LPE InP layers by SIMS analysis, and have seen the reduction of the Si concentration by baking with H$_2$ containing H$_2$O, through the reaction $Si(l) + 2H_2O(g) \rightleftarrows SiO_2(s) + 2H_2(g)$ forming SiO_2 particles in the growth solution, leaving S as the residual donor. However, other workers [192,195] have found that high-purity layers can be obtained without addition of H_2O to the gas stream. It has also been found to be beneficial to bake all the solution components, not just the In [192,195,197], suggesting that significant amounts of impurity are introduced with the solid source materials.

4.3.2 Carrier concentration/doping control

The run-to-run control and the uniformity of carrier concentration are of extreme importance for most devices. In LPE, the carrier concentration in the epitaxial layer is controlled by the addition of impurity species to the growth solution, normally in elemental form. Because, to a first approximation, the segregation of impurity atoms from the solution into the growing layer is an equilibrium process, there is usually a simple relationship between the concentration of dopant in the solid and liquid phases. This means that the carrier concentration control in LPE is basically good. There are complicating factors, some of which were discussed earlier in §3.4:

(i) Sometimes, not all the dopant species are electrically active due to compensation or precipitation. This is particularly the case at higher doping levels.

(ii) The dopant incorporation is dependent on growth rate, temperature and crystal orientation. All these parameters must be controlled.

(iii) At low carrier densities, the effect of random background carriers due to unintentional dopant atoms may lead to poor reproducibility.

(iv) The dopant species may have a vapour pressure at the growth temperature that is sufficiently high for the dopant to be lost at a significant rate from the growth solution.

In spite of these pitfalls, reproducible doping ($\pm 5\%$) can be achieved for most III–V materials in the range $10^{16}–10^{19}$ cm^{-3}. However, reproducibility in the range $< 10^{16}$ cm^{-3} becomes progressively worse as the carrier density falls close to the background level.

There is little data available on the lateral variations of carrier concentration in LPE layers, although Vilms and Garrett [37] have found that the lateral variation over 2 cm^2 LPE layers of GaAs is ± 2–5%. Depthwise grading can occur due to (i) the temperature dependence of the distribution coefficient if the layer is grown over a wide temperature range, (ii) depletion of the dopant species in the solution if $k \gg 1$, or (iii) interdiffusion between layer and substrate (or adjacent epitaxial layers). Keller and Muench [33] found that the carrier concentration in 75 μm thick LPE layers of Zn-doped GaAs varied by $< 10\%$ through the layer, while Vilms and Garrett [37] measured variations of $< \pm 10\%$ through 5 μm thick GaAs LPE layers.

The effects of dopant diffusion can be illustrated by the case of Zn-doped InP where the Zn can diffuse rapidly from the doped LPE layer into the substrate during the growth and can be lost by evaporation from the growth solution. This was shown by Wada *et al* [48] by carrier concentration profiling. Such diffusion processes also cause problems during the growth of GaInAsP/InP DH lasers where Zn, used as a p-dopant for either quaternary or InP layers, can diffuse into adjacent undoped or n-type layers to produce remote junctions. This was clearly shown in the work of Hersee *et al* [95], where a contact resistance profiling technique was used to show the position of the electrical interfaces. Figures 4.7(*a*) and (*b*) show profiles of p$^+$ (Zn-doped) and undoped quaternary layers grown on semi-insulating InP substrates. The diffusion front caused by Zn diffusion into the substrate can be clearly seen in (*a*). This sort of problem has been overcome either by using lower growth temperatures or by using dopant species which diffuse more slowly or are less volatile, e.g. Mg [198] or Be [47] as acceptors in InP instead of Zn.

To summarise, the doping of epitaxial layers in the LPE process can normally produce uniformly doped layers with good run-to-run control over a wide range of doping (10^{16}–10^{19} cm^{-3}) and although problems can occur with particular dopants, these can normally be overcome due to the wide variety of growth techniques, growth temperatures and dopants available in LPE.

4.3.3 Carrier mobility
The carrier mobility, as measured typically by the Hall effect, is an extremely important parameter in the performance of many

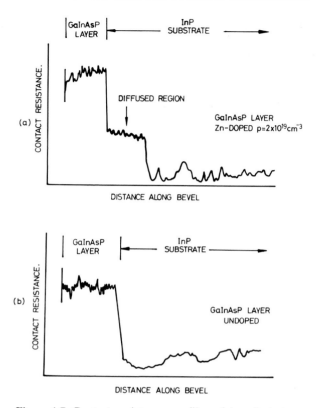

Figure 4.7 Contact resistance profiles of bevelled structures of (*a*) GaInAsP (2×10^{19} Zn doped) layer on undoped InP substrate and (*b*) undoped GaInAsP layer on undoped InP substrate showing the presence of a Zn diffused region in the specimen (*a*) (after Hersee *et al* [95]).

devices, particularly FETs, microwave devices and ICs. The mobility at 77 K or below is normally extremely sensitive to the presence of impurities and other defects in the material. Figure 4.8 shows theoretical curves of μ_{77} versus free electron concentration for n-type GaAs for compensation ratios (N_A/N_D) of 0 and 0.5. Some typical results are also shown for LPE-grown n-type GaAs [77,183,199,200]. These results require compensation ratios in the range $N_A/N_D = 0.4$ to 0.6 to fit to the theoretical curves. Similar ratios have been found in other materials. For example, values of N_A/N_D of 0.4 to 0.6 have been found for Sn and Te-doped InP [62] as shown in figure 4.9. This auto-compensation effect reduces the mobility of the material below that which would be achieved in uncompensated material. In undoped LPE layers there are several reports of 77 K mobilities which closely approach theoretical limits. For example, Otsubo *et al* [199] used controlled baking schedules for the LPE solutions to achieve GaAs layers with N_D – $N_A \sim 4 \times 10^{12}$ cm^{-3} and $\mu_{77} = 2.44 \times 10^5$ cm^2 V^{-1}s^{-1} compared to a predicted lattice-scattering limited value of 2.4×10^5 cm^2 V^{-1}s^{-1} [201].

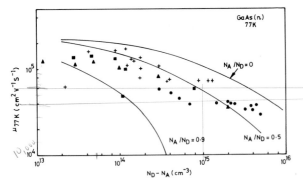

Figure 4.8 77 K Hall mobility (cm^2 V^{-1} s^{-1}) as a function of $(N_D - N_A)$ (cm^{-3}) for n-type GaAs showing theoretical calculations for compensation ratios N_A/N_D of 0.0, 0.5 and 0.9 (after Wolfe *et al* [201]) and experimental results for high-purity LPE-grown GaAs from several sources: (▲) Otsubo *et al* [199,200]; (+) Morkoc and Eastman [77]; (●) Nanishi [183]; (■) Skromme *et al* [182].

For alloy systems, the mobility can be reduced by alloy scattering due to perturbations of the crystal potential such that

$$\frac{1}{\mu_a} = \left[\frac{1}{\mu_0} + \frac{1}{\mu_{alloy}} \right] \qquad (4.2)$$

where μ_0 is the lattice-limited mobility and μ_a is the actual mobility taking account of the alloy scattering.

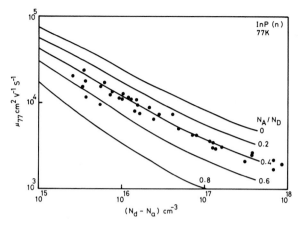

Figure 4.9 77 K Hall mobility (cm^2 V^{-1} s^{-1}) as a function of ($N_D - N_A$) (cm^{-3}) for n-type InP grown by LPE [62].

For $Ga_{0.47}In_{0.53}As$, Pearsall [56] has calculated $\mu_a = 1.4 \times 10^4$ cm^2 $V^{-1}s^{-1}$ at room temperature, using the values $\mu_0 = 2 \times 10^4$ and $\mu_{alloy} = 4.7 \times 10^4$ cm^2 $V^{-1}s^{-1}$. For lattice-matched LPE layers of $Ga_{0.47}In_{0.53}As$ with carrier concentrations at the low 10^{16} cm^{-3} level, values of room temperature Hall mobility of $\mu = 1.2 \times 10^4$ cm^2 $V^{-1}s^{-1}$ have been measured [56] suggesting that the lattice-matched LPE material can be free of strain and defects.

4.3.4 Minority carrier diffusion length

Although not a routine materials assessment for epitaxial layers, the measurement of minority carrier diffusion length (L) is important for several optoelectronic devices such as LEDs, solar cells, IR detectors and photocathodes. Several techniques have been used based on the collection of minority carriers excited by a light spot or electron beam using a Schottky barrier or p/n junction.

One technique uses the electron-beam induced current (EBIC) mode of the SEM, as this is particularly useful for looking across the cleaved edge of an epitaxial structure containing a p/n junction. Several workers have measured the wavelength dependence of the photovoltaic effect to obtain values of diffusion length in epitaxial layers of GaAs [202,203]. The experimental results can be fitted using a simple theory to yield a value of L.

Typical results for LPE-grown n-type GaAs layers are shown in figure 4.10 from the work of Wight *et al* [204]. Some workers have reported extremely long diffusion lengths in high-purity LPE layers. For example, Nelson [205] has reported $L_e = 140 \, \mu$m for an undoped GaAs layer with $p = 1.9 \times 10^{15}$ cm^{-3}, while Ryan and Eberhardt [206] have measured a value of $L_h \sim 200 \, \mu$m in a high-purity ($n < 10^{14}$ cm^{-3}) LPE layer. In studying the hole diffusion length L_h in n-type GaAs grown by LPE, Sekela *et al* [207] found a correlation between solution baking and increased diffusion lengths, implying that a specific defect centre possibly involving oxygen might be responsible for low diffusion lengths.

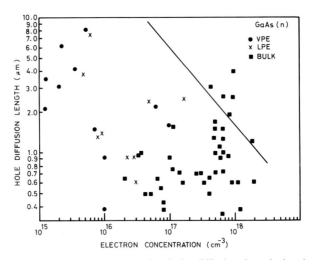

Figure 4.10 Minority carrier hole diffusion length (μm) as a function of electron concentration (cm^{-3}) for n-type GaAs grown by ($\bullet$) VPE, ($\times$) LPE and ($\blacksquare$) melt growth techniques; (—) theoretical curve (after Wight *et al* [204]).

The effect of lattice mismatch in LPE-grown p-type quaternary epitaxial layers on the electron diffusion length (L_e) has been studied by Shirafuji *et al* [208] for In$_x$Ga$_{1-x}$As$_{1-y}$P$_y$ layers

$(x \sim 0.84, \; y \sim 0.63)$. It was found that L_e was almost independent of lattice match between $-4 \times 10^{-3} < \Delta a/a < 3 \times 10^{-3}$, being in the range 0.77 to 1.67 μm. This result was explained by assuming that the carrier lifetime was determined by direct recombination, rather than by the presence of dislocations, interface states or deep electron traps.

The effect of very high dislocation densities on diffusion length has been studied by Booker *et al* [209] in LPE GaAs:Zn layers grown as part of a transmission photocathode structure. In the as-grown material, the electron diffusion length had typically a value of 3 μm whereas after bonding to glass, a process which introduced a number of dislocations, the diffusion length was much lower and correlated with the mean dislocation spacing in the layer.

In summary, the measured minority carrier diffusion lengths in homoepitaxial LPE layers tend to vary from laboratory to laboratory. This may be due partly to measurement errors and partly to varying concentrations of 'killer' centres. In moderately doped material, the values of diffusion length approach those computed from the probability of hole–electron recombination across the bandgap. At low carrier concentrations, however, the results normally fall far short of these theoretically expected values, possibly due to compensation or to the presence of specific centres in the material.

For heteroepitaxial layers with small lattice mismatches, the diffusion length has been found to be independent of mismatch, while it is expected that for large mismatches where misfit dislocations are introduced, the diffusion length may become limited by the mean dislocation spacings.

4.3.5 Deep levels

The presence of deep levels in a material has a great bearing on several of the material parameters already discussed, such as luminescence efficiency and minority carrier lifetime. One of the major advantages of LPE is the low density of both electron and hole traps achieved, due to the fact that (i) LPE is a low-temperature process and thus introduces fewer chemical and structural defects than high-temperature melt growth processes and (ii) LPE growth occurs in conditions of well controlled stoichiometry (usually metal-rich).

In recent years, several sensitive techniques have been developed for measuring the energies (E_T), densities (N_T) and carrier capture cross-sections (σ_T) of electron and hole traps in semiconductors, mostly based on capacitance measurements on Schottky barriers or p–n junctions such as deep level transient spectroscopy (DLTS). It has been found, for example, that LPE-grown GaAs has very low densities ($< 10^{12}$ cm^{-3}) of electron traps [210] in comparison with bulk-, VPE- or MBE-grown material. However, several hole traps have been found in LPE-grown GaAs as shown in table 4.4. Not all these levels are seen in each specimen and in some cases are almost completely absent [211]. Recent work has shown that the density of hole traps can be reduced by controlling the stoichiometry of the layer [199] or by growing using the technique of liquid-phase electro-epitaxy (LPEE) [107]. The density of these hole traps is generally $\sim 10^{14}$ cm^{-3}, which is typically an order of magnitude higher than in VPE-grown GaAs but an order of magnitude less than in bulk-grown GaAs.

Table 4.4 Hole traps commonly seen in LPE GaAs.

Level	Assignment	Reference
(Energy above valence band, eV)		
0.40	Not known	[210–213]
0.45	(Cu_{Ga}–V_{As})	[199,214]
0.50	As-vacancy complex	[215]
0.65	As-vacancy complex	[212,215]
0.72	Fe-related defect complex	[210,211]
0.80	Not known, only seen in p-type undoped material	[212]

Data on deep levels in other III–V materials is rather scarce. In the case of GaP, a clear correlation between the density of the dominant hole trap at ($E_v + 0.75$) eV as measured by photocapacitance and the minority carrier lifetime has been established [223], the density being approximately an order of magnitude lower in LPE material ($\sim 10^{13}$ cm^{-3}) compared with VPE material ($\sim 10^{14}$ cm^{-3}).

Work by Shirafuji *et al* [208] on LPE-grown $Ga_{1-x}In_xAs_{1-y}P_y$ ($x = 0.77$, $y = 0.85$) has shown that there are two dominant electron traps at $(E_c - 0.67)$ and $(E_c - 0.60)$ eV, the densities of which are very sensitive to the lattice mismatch between the layer and the InP substrate, there being a sudden decrease in the density of the $(E_c - 0.67)$ eV level from $\sim 10^{16}$ cm^{-3} at $\Delta a/a = 2 \times 10^{-3}$ to 10^{13} cm^{-3} at zero mismatch.

The deep levels in InP grown by various techniques have been compared by Chiao and Antypas [216]. Four electron traps were regularly seen at 0.58, 0.78, 0.89 and 1.15 eV below the conduction band. All four were seen in bulk InP at densities in the range 10^{13}–10^{14} cm^{-3}. Although only two of the levels, at 0.78 and 1.15 eV below E_c, were regularly seen in LPE material, their densities were quite high (10^{14} and 3×10^{14} cm^{-3} respectively). It has been suggested that the 1.15 eV level may be due to a P-vacancy complex involving Si_{In} donors, and that it may be greatly reduced by pre-baking the growth solution in H_2 containing a few PPM of H_2O [62]. In VPE-grown InP, the two levels at 0.89 and 1.15 eV below E_c were seen but at extremely low concentrations (7.6×10^{12} and 3.8×10^{13} cm^{-3}).

In summary, the incidence of electron and hole traps in LPE-grown layers varies from material to material. Many deep levels in III–V compounds are due to complex centres involving a vacancy and adjacent substitutional impurity, and hence the density of these centres can be lowered by removing the impurities responsible and by reducing the vacancy concentration. In the case of LPE-grown GaAs layers, the residual impurity concentrations achieved are not normally as low as those in VPE material except when extremely careful solution pre-treatments are employed. Also, since LPE growth of III–Vs normally employs group III (metal)-rich growth conditions, this favours the formation of group V vacancy complexes. However, the low growth temperatures used in LPE lead to lower equilibrium concentrations of native defects in comparison with high-temperature melt-growth techniques. The concentrations of deep levels in LPE material are generally intermediate between those in VPE and bulk material. The effect of these differing deep level densities on devices is difficult to assess, since it is the traps with large carrier capture cross-sections which can dominate the recombination process.

4.4 Optical Properties

4.4.1 Luminescent properties

One of the most important assessment techniques for III–V materials grown for optoelectronic devices is the measurement of luminescence spectra. This has been done by three main techniques:

(i) Photoluminescence (PL), where the excitation source is a light spot illuminating the surface of the sample. The excitation density can be varied over a wide range and by using a scanning laser spot, spatial variations across a sample can be measured.

(ii) Cathodoluminescence (CL), where an electron beam, usually in a scanning electron microscope, is used to excite the luminescence. This technique has the advantage of being able to sample very small volumes and can be used in conjunction with other SEM techniques such as electron probe micro-analysis (EPMA) or electron-beam induced conductivity (EBIC).

(iii) Electroluminescence (EL), where minority carrier injection using a p–n junction or Schottky barrier produces the luminescence. This technique is mainly used for measuring LED device performance.

Since photoluminescence has been the most widely used technique, most of the following discussion will concentrate on this. There are several types of luminescence measurement, as discussed in the following sections.

Luminescence spectra

A luminescence spectrum can yield information on the concentration (and in some cases the identity) of impurities in an epitaxial layer and on the concentration and possible structure of complex centres involving native defects and foreign impurities. At room temperature, a broad near-band-edge peak is typically seen, which can be used as a rapid assessment of the bandgap of the material. An example is shown in figure 4.1(b) for the case of $In_{1-x}Al_xP$ using CL at 300 K. With the recent interest in ternary and quaternary alloys, both PL and CL at room temperature have been widely used techniques for rapid bandgap assessment and for compositional mapping of epitaxial wafers.

At lower temperatures ($<$ 100 K), several near-band-edge peaks can be resolved. For example, figure 4.11 shows PL spectra for low n-type GaAs [217] grown by LPE and VPE. The peaks have been assigned as follows:

A   (1.518 eV) exciton recombination and donor-to-valance band transitions

B   (1.49–1.50 eV) conduction band to acceptor transitions

C   (1.35 eV) As-vacancy/acceptor complex

D   (1.20 eV) Ga-vacancy/donor complex

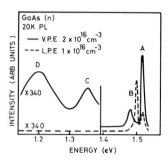

Figure 4.11 Photoluminescence spectra at 20 K of LPE- and VPE-grown low n-type GaAs layers (after Williams and Blacknall [217]).

The main points to notice are (i) the absence of the deeper levels C, D in LPE-grown material and (ii) the slightly different positions of peak B for the LPE and VPE material due to the differing acceptor levels involved in each case.

At even lower temperatures (2 to 20 K), the structure of the excitonic region (peak A) becomes resolved in high-purity samples as shown in figure 4.12 from the work of Skromme *et al* [182]. The band-to-acceptor region (peak B) at low temperatures is resolved into free-to-bound and donor–acceptor peaks. Also the so-called two-hole replicas of the acceptor–bound exciton transitions can be resolved in this region and have proved to be useful as 'fingerprints' for the identification of residual acceptors in GaAs. Ashen *et al* [218] have used these replicas to show that the residual acceptor in VPE GaAs was Zn while in their LPE GaAs there were three residual acceptors, C,

Si and Ge. In more recent work on high-purity LPE GaAs from several laboratories, Skromme *et al* [182] have identified C and Si as the most common residual acceptors in LPE GaAs, and using photothermal ionisation measurements have shown that S is by far the most common background donor.

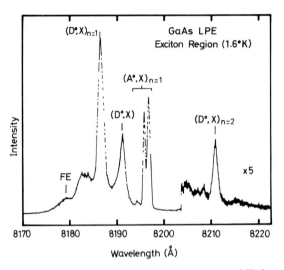

Figure 4.12 Photoluminescence spectrum at 1.6 K from high-purity LPE-grown GaAs layer showing the excitonic structure in the near-band-edge region (after Skromme *et al* [182]).

Relatively little work has been reported on the use of luminescence to investigate the defect structure of alloys. Fukui *et al* [219] have used PL and EL to investigate the deep-level emission in LPE-grown InGaAsP/InP double-heterostructure lasers, which causes broadened emission peaks. The luminescence of LPE $Ga_{1-x}Al_xAs$ alloys has been studied by several workers [220,221]. Rogulin *et al* [220] found that in addition to the near-band-edge emission which, as expected, shifts to higher energy as x increases, there is a deep level peak at ~ 1.2 eV (cf GaAs) whose intensity increases with x but whose peak position remains constant.

Luminescence intensity measurements

For the case of LED materials, information is often required on the efficiency of a particular luminescent transition in competition with other radiative and/or non-radiative processes, as a function of doping and other crystal growth variables. For example, in the case of GaP:N, the efficiency of the near-band-edge green emission due to recombination of excitons bound to the isoelectronic nitrogen centre is reduced by (i) competing radiative processes through deeper levels, e.g. the red emission due to recombination of excitons at nearest-neighbour Zn–O complexes and (ii) the effect of non-radiative Auger recombination.

Luminescence measurements can be used to compare the intensity of the luminescent transition of interest with that of a competing transition or with a standard sample. This type of measurement is often done in conjunction with a separate assessment (e.g. by photocapacitance) of deep level densities. For example, Dapkus *et al* [222] have studied the external PL quantum efficiency of LPE-grown GaP:N layers as a function of donor and acceptor concentration. The simultaneous measurement of PL decay time enabled the authors to confirm a model for the kinetics of the recombination in this material, which is important to understand the reasons for the low radiative efficiency.

The spatial variation of CL intensity obtained by rastering the electron beam in an SEM across the sample surface can be a useful way of correlating luminescence efficiency with defects located or identified in the material by other techniques, e.g. etching or TEM. Similar mapping can be carried out by PL obtained by the use of a scanned focused laser beam as the excitation source.

4.4.2 *Minority carrier lifetime measurements*

The other major area where luminescence measurements have provided data on epitaxial layers is in the measurement of minority carrier lifetimes from luminescence decay characteristics. This has generally been done by using short pulsed laser excitations (PL) or by fast modulation of the electron beam in the SEM (CL). Such measurements can be correlated with measurements of luminescence efficiency or the concentration of particular deep levels obtained by photocapacitance or deep level transient spectroscopy (DLTS). For example, Hamilton *et al* [223] were able, by measuring relative CL efficiency and minority carrier lifetimes in GaP epilayers, to show a correlation between these parameters and the concentration of a hole trap at $(E_v + 0.75)$ eV measured by photocapacitance. This trap, thought to be related to P-vacancies, was found at a concentration of $\sim 10^{14}$ cm^{-3} in VPE-grown layers and 10^{13} cm^{-3} in LPE layers of GaP.

In summary, the internal luminescent efficiency of LPE-grown layers of III–V binaries is normally high due to the low concentrations of deep-level carrier traps, low dislocation densities and the uniformity of the layers.

4.5 Crystal Perfection

In this context, crystal perfection will encompass not only dislocations and stacking faults, but also the layer surface topography. These factors have an important impact on device performance and device yield.

4.5.1 Dislocations

The various mechanisms which can produce dislocations in epitaxial layers of III–Vs have been thoroughly reviewed by Booker *et al* [209]. Figure 4.13 shows some of the dislocation behaviours seen in epitaxial structures.

The major causes of dislocations in LPE layers are:

(i) *Propagation of substrate dislocations*: although a high proportion of the substrate dislocations can propagate during epitaxial growth, this process can be reduced by controlling the stresses bending the dislocations into the interface plane (figure 4.13(*a*)). The force acting on the dislocation F_m due to the mismatch is proportional to the lattice mismatch $(\Delta a/a)$ and the layer thickness. Rozgonyi *et al* [224] demonstrated this mechanism by studying the effect of increasing the mismatch in the LPE growth of $Ga_{1-x}Al_xAs_{1-y}P_y$ on GaAs substrates when y is varied from 0 to 0.05 and the layer thickness is increased. It

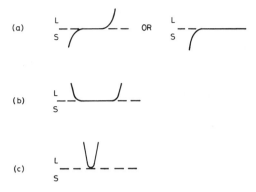

Figure 4.13 Various types of dislocation behaviour in epitaxial structures, where S, L represent the substrate and layer respectively.

has been reported that even in the homoepitaxial LPE growth of GaAs on GaAs substrates, marked reductions of the dislocation density in the layer compared to the substrate can be achieved [225]. This was ascribed to the effect of small thermal stresses at the interface during epitaxial growth.

(ii) *Differential contraction*: in heteroepitaxial growth, even if lattice matching is achieved at the growth temperature, there will always be some contribution from differential contraction as the grown structure cools down to room temperature, since the expansion coefficients of layer and substrate are rarely the same. This is the case for the growth of several ternary and quaternary systems, e.g. (GaAl)As on GaAs, (Ga, In)(As, P) on InP. As the structure cools, the stress caused by differential contraction can be relieved by tetragonal distortion of the unit cell of the epitaxial layer and some degree of bending of the structure which depends on the relative thicknesses of the layer and substrate. If the surface stress becomes too high, however, and depending on the stress rate, dislocations may be generated at the surface (figure 4.13(*b*)) and move down to the interface. This latter process has been thoroughly studied by TEM investigations, particularly in the (GaAl)As/GaAs system [209]. The tetragonal distortion of the unit cell has been well documented for the (GaAl)As/GaAs [226] and (Ga, In)(As, P)/InP systems [227].

(iii) *Lattice mismatch at the growth temperature*: when the epitaxial layer has a lattice mismatch with the substrate of greater than $\sim 10^{-4}$ at the growth temperature, dislocations are likely to be found as the layer thickness reaches a critical value, probably by the surface generation mechanism mentioned earlier. These dislocations glide to the substrate–layer interface and produce a dislocation network in or near the interface plane. This type of ·behaviour has been observed frequently in LPE-grown heterostructures at high mismatches. For example, Oe *at al* [227] have established that in the growth of (Ga, In)(As, P) layers on InP at 645 °C, no misfit dislocations are produced if the mismatch $|\Delta a/a|$ is $< 5 \times 10^{-3}$. The extremely good composition control of alloy composition achieved in LPE has enabled this source of dislocations to be eliminated in this and other such systems. This has important consequences for several device applications.

When the lattice mismatch is very large $(> 10^{-2})$, then dislocations are spontaneously produced at the interface from the earliest stages of growth often in pairs as shown in figure 4.13(*c*). The threading dislocation density in this case is usually very large and would not be suitable for device applications.

4.5.2 Stacking faults

The presence of stacking faults can be readily seen on the as-grown surface of an LPE layer (figure 4.14). Almost invariably, the stacking faults are initiated at the substrate surface due to the presence of contamination in the form of either foreign particles· or native oxide. These problems can be solved in several ways:

(i) by means of *in situ* etching back of the substrate immediately before the commencement of growth with a wash melt (see §4.5.3)

(ii) by careful substrate surface preparation to remove particulate contamination and surface oxides, and

(iii) by pre-growth baking of the substrate in the H_2 ambient in the LPE kit.

4.5.3 Layer surface topography

There are several commonly encountered surface features of LPE layers:

Figure 4.14 Stacking faults on the surface of LPE (GaAl)As layer grown on to (111)B GaP (© Crown Copyright, HMSO, 1990).

(i) *Terrace pattern*. This is probably the most distinctive surface feature of LPE-grown material. It is also sometimes known as the sea-shore or ripple pattern. The relationship of this pattern to the deviation of the substrate surface from a low-index plane was discussed earlier in §1.4.1. The key to the removal or reduction of this effect is careful substrate orientation, which in turn means accurate x-ray orientation of the substrate ingot and great care in the subsequent cutting and polishing stages to ensure that the substrate orientation is maintained from slice to slice and over the whole area of each slice. The other important parameter is the supersaturation of the growth solution which controls the nucleation density.

(ii) *Pinholes*. These are holes which go right through the layer to the substrate and are caused by non-wetting of the substrate surface or an inhibition to nucleation caused by contamination. This problem can be overcome by better substrate preparation or by the use of *in situ* wash melts to etch off the top few microns of the substrate surface in the LPE kit immediately before growth [27].

(iii) *Indentations or pits*. These can be caused either by a

pinhole defect which has partially filled in during the growth, or
to replication of the indentations on the substrate–layer interface due to localised etch-back of the substrate. (See
figure 4.15(*a*).)

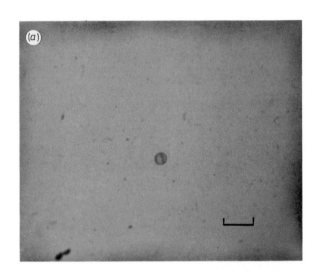

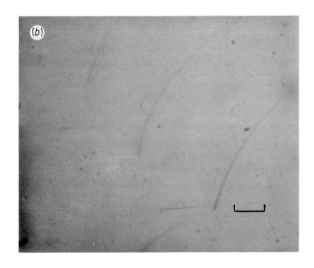

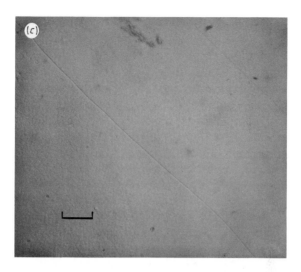

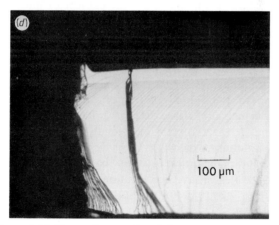

Figure 4.15 Photomicrographs of features on LPE-grown layers: (*a*) small crater (50 μm marker); (*b*) cusp defects (50 μm marker); (*c*) meniscus lines (50 μm marker); (*d*) edge growth seen in cross-section (100 μm marker) (© Crown Copyright, HMSO, 1990).

(iv) *Cusp defects.* It has been postulated that these are caused when the lateral movement of growth steps across the layer surface is pinned at certain positions by small surface protruberances or depressions [228]. (See figure 4.15(*b*).)

(v) *Meniscus lines*. These fine lines on the layer surface are caused by the movement of the growth solution on or off the substrate and are particularly noticeable in the sliding boat technique. In cross-section they consist of an S-shaped 'ridge and furrow' of typical height 100–500 Å (figure 4.15(*c*)). They are thought to be caused by the 'stick–slip' motion of the solution, whereby the solution dwells for a short time at certain positions. Small *et al* [229] have related this effect to the unique three-phase (L–V–S) boundary where the solution surface is in contact with the substrate, which could affect the local equilibrium temperature. This feature of LPE surfaces is generally not a problem and can be removed largely by ensuring rapid solution removal. This can now be done reproducibly with stepper-motor-driven push-rod assemblies.

(vi) *Edge growth*. This is caused by an enhanced growth rate round the perimeter of the substrate, as shown in figure 4.15(*d*). The exact cause of this has been the subject of much research, and several practical solutions have been proposed. The main theories have been that (a) the high index planes revealed at the curved rim of the substrate have higher growth rates than the low index substrate surface, (b) there is an increased heat loss from the graphite walls of the solution well, (c) that the equilibrium shape of the liquid–solid interface is perturbed by the different balance of surface tension forces at the edge of the solution and (d) that three-dimensional solute diffusion occurs from the solution which lies outside of the wetted region [122,230]. The main practical steps that can be taken to avoid this effect are (a) to have the solution smaller in lateral dimensions than the substrate, (b) to use thin solutions and (c) to use slow growth rates.

4.5.4 Stoichiometry control

Several workers have reported on the control of the stoichiometry of LPE-grown III–V material by means of the vapour pressure of the group V species [30,81,231–233]. This has been applied to GaP, GaAs, (Ga, Al)As, InP and (Ga, In) (As, P). The phase rule for a two-component system (GaAs, GaP) ($c = 2$) would suggest that when liquid, solid and vapour are all at equilibrium ($p = 3$), the number of degrees of freedom f should be $f = c - p + 2 = 1$. Once the temperature has been selected, the composition of all the phases should thus be fixed. Nishizawa and Okuno [232] have made the point that when the deviation of stoichiometry of the material δ is taken into account, i.e. $GaAs_{1-\delta}$, a new degree of freedom is introduced whereby at any temperatures T, δ can be changed within the solidus existence limits by changing the partial pressure of the group V species in the vapour phase. By carrying out LPE growth of GaP and GaAs by the constant-temperature-gradient method (see §2.3.5) with separate control of P or As vapour pressure using a heated elemental source, Nishizawa and Okuno [232] were able to show that there was an optimum vapour pressure where:

(i) the defect density in the layers is much reduced
(ii) the surface topography is smoothest
(iii) the carrier concentration was a minimum
(iv) the lattice constant of the GaAs layers was a minimum
(v) the deep level concentrations were minimised.

In more recent work, Novotny *et al* [231] have grown GaAs/(GaAl)As double heterostructures (DH) for laser applications in a conventional sliding-boat system under controlled arsenic vapour pressure. The effects of the arsenic vapour pressure were to produce:

(i) improved surface morphology
(ii) flatter interfaces at the heterojunctions
(iii) sharper doping profiles
(iv) laser threshold current reduced by a factor of two.

In the case of InP and InGaAsP layer growth, the use of a phosphorus overpressure produced by introducing PH_3 or elemental phosphorus into the H_2 ambient gas has been investigated by Astles *et al* [46], Takahashi and Nagai [30] and Keramides *et al* [233]. The effects of having a phosphorus overpressure were to:

(i) prevent degradation of the InP substrates due to loss of P before growth;
(ii) prevent a similar loss of P from the grown layers during the post-growth cool-down;
(iii) reduce the PL intensity of the 0.99 eV band associated with P-vacancies;

(iv) increase the PL intensity of the (Ga, In)(As, P) active layer in a DH laser structure;

(v) reduce the threshold current of the DH lasers.

There are no direct measurements of the effects of these group V overpressures on deep level concentrations in the layers in any of the papers mentioned above but the results quoted point clearly to the fact there are marked effects of the gas-phase composition on the properties of LPE material.

4.6 Layer Thickness Control

In modern device structures, there is an increasing emphasis on thin layer structures and it is in this range (10 nm to 1 μm) where LPE has difficulty with thickness control. From the discussion earlier in §3.2, it can be appreciated that the primary growth parameters which affect the thickness of an LPE layer are the supercooling ΔT_s of the growth solution, the cooling rate R, and the growth time, t, as indicated in the general equation (3.20) which can be written as

$$d(t) = \frac{\text{constant}}{m} \left(2\Delta T_s t^{1/2} + \frac{4}{3} R t^{3/2} \right) \qquad (4.3)$$

where m is the slope of the liquidus line, $\partial T/\partial C$, which is a function of temperature.

Of these three parameters the most difficult to control is the supercooling ΔT_s, given by

$$\Delta T_s = T_0 - T_g \qquad (4.4)$$

where T_0 is the equilibrium liquidus or saturation temperature of the solution and T_g is the starting growth temperature of the LPE run, i.e. the temperature at which the liquid phase and the substrate are brought into contact.

There are four primary techniques of LPE used for the growth of thin layer structures, all based on the sliding-boat apparatus and all with associated control and reproducibility problems as discussed below.

(i) *Near-equilibrium growth.* In this case the solution composition and hence T_0 is determined by accurate initial weighing of the solution components. No extra source material is added and the solution is not contacted to a source seed or dummy substrate prior to contact with the actual growth. Very small initial supercooling is used and hence extremely accurate weighing of the solution components is required. For example, Besomi *et al* [234] have used the near-equilibrium technique for the lattice-matched growth of InGaAsP ($\lambda = 1.3 \ \mu$m) on to InP (100) substrates and calculated that to obtain $\Delta T_s = 0.1 \ ^\circ$C reproducibly, the precision of weighing the InP for the quaternary solution needs to be $\pm 0.07\%$, and for the InAs and GaAs constituents, $\pm 0.1\%$. Another problem with this 'dead reckoning' technique is the loss of a volatile constituent from the solution during the pre-growth heating stages. For example, Cook *et al* [120] have measured a decrease in liquidus temperature in InGaAsP solutions of 0.71 $^\circ$C h^{-1} when heated at 670 $^\circ$C and 0.34 $^\circ$C h^{-1} at 640 $^\circ$C, due to loss of P by evaporation. However, this should be reduced by the use of tight-fitting solution caps. In spite of these difficulties, several authors have favoured this technique for the growth of the thin active layers in InGaAsP/InP double-heterostructure layers [234,83]. A variation of this technique is the use of a saturation slice. For example, in the growth of GaInAsP, a starting ternary mixture of In–Ga–As is put in contact with an InP disc for several hours to saturate the solution fully. The orientation of the saturation disk should match that of the substrate [83] to get the best results, because of the differing equilibrium solid compositions on different orientations.

(ii) *Two-phase solution technique.* This technique relies on the presence of solid material in the solution during the ramp-cooling growth process, either in the form of excess saturation material or as a crust of grown equilibrium solid. This has the advantage of reducing the need for very accurate weighing of the solution components and removes a certain amount of excess supersaturation in the growth solution [85,235,236]. However, due to the deposition of material on the excess solid material and the consequent uncertainty in the solution supersaturation, it is difficult to get very reproducible thin layer growth by this technique. For example, the results of Nelson *et al* [236] show run-to-run variations in layer thickness of a factor of two for fixed growth times in the thickness range 0.1–0.5 μm.

(iii) *Source-seed technique.* Another approach to controlling supersaturation in the growth solution is the source-seed technique whereby the growth solution is positioned over a sacrificial polycrystalline or single crystal slice of the same material as the substrate immediately prior to contacting the solution with the substrate. This is designed to relieve any excess supersaturation in the solution. However, Besomi *et al* [234] obtained poor thickness reproducibility with this technique, and attributed this to a tumbling of the growth solution when it was moved from the source to the substrate causing high supersaturation liquid from the top of the solution to move to the bottom near the substrate.

(iv) *Super-cooled/step-cooled growth technique.* Both of these methods use an initial supersaturation of the solution of typically 5–10 °C. In situations of heteroepitaxy where a previously grown layer is susceptible to etch-back by the oncoming growth solution, or where there is a lattice mismatch between the growing layer and the substrate, such large supersaturations may be vital to grow the desired structure. However, once the nucleation has been accomplished, further growth may be rapid, leading to difficulty in controlling the layer thickness. Although the step-cooled (constant-temperature) growth technique is claimed by Feng *et al* [86] to produce uniform composition layers while the techniques involving temperature ramps give graded layers, such grading is not a problem in sub-micron thickness layers.

For the growth of very thin layers for quantum-well devices (100–1000 Å) extremely accurate control of growth time 't' as well as the supersaturation ΔT_s is required. In this regard, the use of motorised movements under computer control for linear or rotary sliders has been a key development [237–241]. With manually operated slider movements, the growth times cannot be controlled to better than 1 s. Thus to achieve better than ±10% run-to-run variation, growth times of 10 s or longer are required. Also the transfer velocity between wells in a manually operated system cannot be greater than $\sim 20 \, \text{mm s}^{-1}$, whereas for a motorised system velocities of $700 \, \text{mm s}^{-1}$ are possible. Using an automated rotary slider, Rezek *et al* [237] have achieved growth times as short as 18 ms and, by using a relatively large supercooling ΔT_s of 10 °C, were able to grow 140 Å thick layers of GaInAsP. At growth times of < 100 ms, a transient growth mechanism has been found to operate [238] rather than the normal diffusion-limited growth process. This produces a different initial composition from that obtained in longer period growths. This problem will be discussed in more detail in §4.7. To get over this problem, Rezek *et al* [238] have grown stacks of these initial thin transient-growth layers of GaInAsP lattice matched to InP. The other approach used by Capella *et al* [239] is to reduce the supersaturation ΔT_s to 2 °C and to reduce the growth temperature (which increases the value of *m* in equation (4.3) and hence reduces the value of *d* for when all other parameters are equal). This enabled them to grow ~ 100 Å thick layers of InGaAsP with growth times of 2 s. Similarly, Zwicknagel *et al* [240] have grown p–n–p–n GaAs multilayers with individual layer thicknesses as low as 200 Å by using lower-than-normal growth temperatures of 630 °C or 550 °C and ramp-cooling rates of 0.5–0.75 °C min^{-1}, the growth rates being in the range 500–2000 Å min^{-1}.

In summary, it can be seen that by proper choice of the parameters *m*, ΔT_s and *t* in equation (4.3), thin layers down to ~ 100 Å can be grown by conventional LPE techniques such as the rotary and linear sliding-boat systems.

4.7 Abrupt Junctions

The ability to grow abrupt heterojunctions or doped homojunctions is an important facet of any epitaxial growth technology. In the case of liquid-phase epitaxy this is achieved by having a variety of growth solutions in the growth system, one for each of layers in the desired structure, which are sequentially brought into contact with the substrate. The abruptness of the junctions grown by LPE, i.e. the distance over which the change in composition or dopant concentration takes place, is mainly determined by four factors: (i) substrate instability effects, (ii) transient growth, (iii) cross-contamination problems and (iv) diffusion.

The phenomenon of the instability of a binary substrate or a previously grown binary, ternary or quaternary layer in the oncoming growth solution has been discussed earlier in §3.3.6. The key to overcoming this problem is the use of highly supersaturated growth solutions to ensure rapid nucleation of the new layer before the previously grown layer is significantly etched-back. Complex boat designs and growth procedures have been used to achieve this high supersaturation in multilayer growths. For example, a particularly difficult etch-back situation is found in the growth of InP or a InGaAsP quaternary layer with high P content onto (InGa)As layers. Nakajima *et al* [242] have established the conditions of supercooling (ΔT_s) and cooling rate (R) as a function of growth temperature to avoid dissolution of $In_{0.53}Ga_{0.47}As$ when InP is grown on top of it, using a (111)A InP substrate. At low temperatures lower supercoolings and cooling rates were required to avoid the melt-back problem. Matsumoto *et al* [243] have established a technique using removable sources on top of the solutions to initially establish saturation (source in place) and then later to allow supercooling (sources removed). This allowed them to grow InGaAsP onto InGaAs without melt-back on the (100) orientation. Fast cooling rates ($\sim 10\,°C\,min^{-1}$) were used to produce the supersaturation rapidly and hence reduce the possibility of spontaneous nucleation in the solution. The effect that this melt-back can have on interface abruptness has been measured by Cook *et al* [244] using Auger depth profiling of InGaAs/InP, InGaAs/InGaAsP and InGaAsP/InGaAs interfaces. In the former two cases, the interface was ~ 90 Å while in the latter case the interface was significanly broader (> 200 Å), and this has been attributed by these workers to the effects of melt-back of the InGaAs layer. Feng *et al* [245] have shown in similar work on InGaAsP/InP/InGaAsP structures, grown on InP substrates, that the chemical transition width of the InP/InGaAsP interface increases with increasing mismatch, which is what would be expected as the rate of nucleation of the InP layer would be reduced allowing more time for melt-back to occur.

A transient growth effect other than that associated with the initial melt-back phenomenon just discussed can occur. This has been studied in detail by Brunemeier *et al* [246]. It manifests itself as a difference in composition of the growing layer at very short growth times < 100 ms compared to that found in normal longer-period LPE growth experiments. By growing a stack of each of the 'transient' layers, these workers were able to measure the difference in bandgap (E_g) and lattice mismatch ($\Delta a/a$), between these stacks and normal long-period LPE layers. These differences are plotted in figure 4.16 as a function of the alloy composition parameter y in $Ga_xIn_{1-x}As_yP_{1-y}$, from which it can be seen that the bandgap of the transient growth layer is always less than that of a normal layer. This effect is obviously not due to melt-back, and has been attributed to the convection produced in the solution when it is moved onto the substrate [246]. This problem can be overcome by actually using a stack of such 'transient' layers in the device structure. This effect has not been seen by all workers, which is not surprising as (i) it may well be masked by melt-back effects and (ii) the slider velocity may well be an important factor which will vary from one laboratory to another.

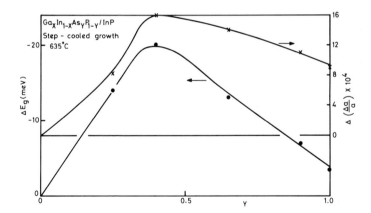

Figure 4.16 Differences between transient growth and normal LPE growth in room-temperature bandgap, ΔE_g (meV) (●), and lattice mismatch to InP, $\Delta(\Delta a/a)$ (×) as a function of y for near-lattice-matched growth of $Ga_xIn_{1-x}As_yP_{1-y}$ on InP (after Brunemeier [246]).

The third practical problem affecting interface abruptness is that of cross-contamination of growth solutions. This can occur by transfer of small amounts of one growth solution to an adjacent one in a slider boat, due to poor wipe-off of the solution from the substrate surface [237]. This can obviously cause changes in the growth rates and compositions of the layers. As the picked-up solution mixes in with the oncoming solution, gradients of composition can occur in the growing layer as has been observed in the growth of GaAs/(GaAl)As structures [247]. This problem can be eliminated by careful control of the wiping clearance in the boat and by using an automated slider movement [248].

The effect of temperature on interface abruptness is chiefly through the increase of diffusion coefficients of matrix or dopant species with temperature. In general, the interdiffusion of matrix species is not a serious problem as the diffusion coefficients are low, e.g. D_{Al} in GaAs at 800 °C is $< 10^{-20}$ cm^2 s^{-1} [249] and the other effects mentioned above are dominant. Although the growth temperatures in LPE are often higher than for VPE or MBE, the growth rates are higher and hence growth times are shorter, thus minimising the diffusion time.

More problems have been encountered with impurity diffusion across junctions, particularly with the group IIB dopants, Zn and Cd. For instance, in the case of Zn in InP, the diffusion coefficient at a typical growth temperature of 640 °C has been estimated as $D_{Zn} \sim 3 \times 10^{-11}$ cm^2 s^{-1} [95], and this leads to misplaced p–n junctions in structures involving InP, GaInAs and GaInAsP layers [95,250]. Because of this, several alternative dopants have been investigated such as InP:Be [47], InGaAsP:Mg [251], InGaAs:Mg [252], InGaAs:Mn [253] and InGaAsP:Mn [254] whose diffusion coefficients are lower than Zn or Cd. For instance, Alavi *et al* [251] have studied the abruptness of p$^+$–n InGaAs/InP heterojunctions using SIMS depth profiling and have shown that for 700 nm thick layers grown at 600 °C in 5 s, the dopant concentration in the InP substrate falls by two orders of magnitude within 52 nm for Mg doping compared with 155 nm for Zn, and one order of magnitude within 32 and 90 nm respectively. These results highlight a great advantage of LPE, namely the ability to vary the dopant species easily.

In summary, it can be seen that there are practical difficulties to be overcome in using LPE to grow abrupt junctions. However, by a suitable choice of growth conditions, such as growth temperature, growth time, supercooling, slider velocity and dopant species, interface abruptness of ~ 10 nm can be achieved in terms of alloy composition changes. This is not as good as can be achieved by MBE or VPE techniques. For example, in GaAs–(AlGa)As heterojunctions grown by LPE (800 °C), VPE (750 °C) and MBE (586 °C), the 10–90% variation distance of Al concentration has been measured as 10, 2 and 1.5 nm respectively [255]. More recent results on MOVPE-grown GaAs–(GaAl)As structures [256] have demonstrated that layers as thin as 2 nm can be grown and hence the transition region width between the layers must be significantly less than this. As well as being unable to grow layers in the 1–5 monolayer range as can be achieved by MBE and VPE techniques, LPE growth runs into problems of thickness reproducibility in a multilayer sequence. This is due to the decreasing supersaturation in the growth solution during the growth sequence, and the fact that in the pauses between growth periods when the particular growth solution is not in contact with the substrate, the solute profile may relax to some extent. Thus, the growth periods need to be extended as the layer sequence builds up in order to keep the layer thickness constant [257]. Thus it can be appreciated that while quantum well and superlattice structures have been grown by LPE, the technique does not readily lend itself to the growth of such structures.

4.8 Selective Area Growth and Growth on Structured Substrates

The fact that LPE growth occurs under near-equilibrium conditions with low supersaturation means that the nucleation is very sensitive to the lattice parameter of the substrate surface and the growth rate is strongly influenced by the crystallographic orientation of the substrate surface. These effects have been used (i) to obtain localised growth in windows on the substrate surface and (ii) to grow on non-planar substrates with ribs or channels produced by preferential chemical etching. The latter effect has proved to be extremely useful in fabricating novel

semiconductor laser structures such as the buried heterostructure laser [258] and the inverted-rib waveguide laser [259].
These will be discussed in more detail in Chapter 5.

The simplest case of selective area growth is where a planar
substrate surface is coated with a masking film such as pyrolytic
or plasma-deposited SiO_2 or Si_3N_4. Windows can then be
opened up in this film using standard photolithographic techniques. When used as a substrate for LPE, growth occurs only in
the window areas with no nucleation on the mask material. In
this way isolated mesas can be grown.

However, the growth on non-planar substrates, normally
without a masking material present, has proved to be of much
greater interest. The structuring normally consists of channels or
ribs produced by preferential etching. It is important to produce
the correct crystallographic plane on the walls of the rib or
channel. This is influenced by the etch solution, the orientation
of the etch windows on the substrate surface and the nature of
the mask material. Figure 4.17, taken from the work of Turley
and   Greene   [260],   shows   the   ribs   produced   by
bromine/methanol etching on (100) InP, showing the difference
in rib shape obtained when the alignment of the SiO_2 mask is in
the [011] or [011] direction. In either case the slow etching
{111}A planes form the walls of the rib. By using other etches
such as hydrochloric acid/phosphoric acid mixtures, ribs with
walls consisting of other crystallographic planes such as {011} or
{111}B can be produced.

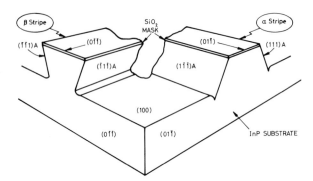

Figure 4.17 Channels and ribs produced by etching of
masked ⟨100⟩ InP substrates with bromine-methanol
(after Turley and Greene [260]).

In considering LPE growth onto such structured substrates, there are two main effects.

(i) Firstly, the growth rate in LPE is affected by the topography of the substrate surface, such that growth is faster over a concave surface than a convex one. This leads to a smoothing out of surface features and it has been suggested that this is due to the excess chemical potential of a curved liquid–solid interface, which can be minimised by a return to planarity [261]. This effect seems to be the dominant one in the growth of GaAs/(GaAl)As on structured substrates.

(ii) The other consideration is the effect of crystallographic orientation on the nucleation and growth kinetics. In the growth of InP and lattice-matched (GaIn)As and GaInAsP alloys on structured InP substrates, this seems to be the major effect. As a result of their work on channelled InP substrates, Chand *et al* [127] listed the ease of nucleation for each material as follows:

InP $\{100\} > \{111\}B \gg \{111\}A$

$In_{0.53}Ga_{0.47}As$ $\{100\} \gg \{111\}A > \{111\}B$

$In_{0.75}Ga_{0.25}As_{0.55}P_{0.45}$ $\{100\} \gg \{111\}B > \{111\}A.$

The relative ease of nucleation of $\{100\}$ and $\{111\}$ faces has been explained [127,262] using the model first put forward by Sangster [10] and discussed earlier in §1.4.1. This model predicts that attachment of group III or V atoms to the $\{100\}$ surface can occur equally easily with no significant energy barrier. However, on a $\{111\}$ surface a stable nucleation cluster needs to consist of three atoms of the opposite type and one atom of the same type as the substrate surface (group III or group V rich). To explain the differences between $\{111\}A$ and $\{111\}B$, the relative probabilities of attachment of the species necessary to nucleate growth on the two faces have been invoked [127,260]. For instance, in the case of InP growth, the use of an In-rich solution makes the attachment of In atoms or (In_3P) clusters to the (111)B face statistically more probable than the reverse situation of P atoms or (P_3In) clusters to the (111)A face.

Finally, a secondary effect of such selective growth behaviour can occur whereby the preferential growth in two adjacent channels can cause a local reduction in solute concentration (i.e. supersaturation) such that growth on the planar surface region between the channels is prevented or very much reduced. This effect can be exploited in laser device structures, as will be described in Chapter 5.

4.9 Composition Control

In the growth of ternary and quaternary alloys, the composition of the alloy is an extremely important parameter to be able to control since it will affect the bandgap of the material, which in many devices needs to be within tightly controlled limits. In most alloy systems, the composition of the alloy also needs to be carefully controlled to obtain the lattice-matched growth which is needed to minimise the density of detrimental defects in the layer.

To a first approximation, the alloy composition can be predicted and controlled from a knowledge of the phase diagram, whereby the primary control parameters are the solution composition and the growth temperature. The other parameters which can have a significant effect are the supercooling (ΔT_s) and the substrate orientation. The effects of these parameters have been studied in some detail for the GaInAsP/InP system. Figure 4.18 shows the results of Hsieh *et al* [96] who used the same liquid composition to grow on (100) and (111)B InP substrates as a function of supercooling and measured the resulting alloy composition. While the (100) orientation is relatively insensitive to changes in ΔT_s, the concentrations of P and Ga in the layers grown on (111)B fall markedly as ΔT_s increases. Even at $\Delta T_s = 0$ a consistent difference in the distribution coefficient of P between the two orientations is found. Because of this result growth on (100) substrates is preferred, as the effect of ΔT_s on composition is negligible for $\Delta T_s < 6\,°C$. De Cremoux [263] has calculated the expected variation in ($\Delta a/a$) and energy gap wavelength (λ) as a function of supersaturation ΔT as

$$\frac{\delta(\Delta a/a)}{\delta(\Delta T)} \sim 3.2 \times 10^{-4}\ °C^{-1}.$$

and

$$\frac{\delta\lambda}{\delta(\Delta T)} \sim 1.3 \times 10^{-2} \ \mu m \ {}^\circ C^{-1}.$$

If the supersaturation of the solution can be controlled to 0.5 °C (a realistic figure) then run-to-run variations of $\Delta(\Delta a/a) \sim 1.6 \times 10^{-4}$ and $\Delta\lambda \sim 7 \times 10^{-3} \ \mu m$ (7 nm) should be achievable. The figure for $\Delta a/a$ is well within the limit of $\pm 5 \times 10^{-4}$ required to avoid introducing misfit dislocations in this alloy system [56], while variations in λ of up to 25 nm are acceptable for laser applications [234].

In terms of the reproducibility of solution composition, Besomi *et al* [234] have calculated that to achieve run-to-run reproducibility of $\Delta a/a$ of $\pm 5 \times 10^{-4}$ for GaInAsP layers on InP substrates, the most demanding requirement is to weigh the InP source material to 0.3%. Over a series of 11 consecutive wafers, they actually measured $\Delta a/a$ values in the range $< 1 \times 10^{-5}$ to $\sim 2 \times 10^{-4}$.

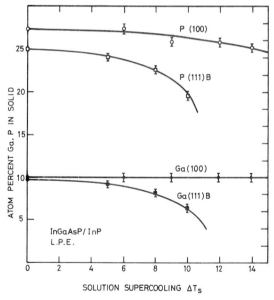

Figure 4.18 Variation of composition of InGaAsP layers grown on $\langle 100 \rangle$ and $\langle 111 \rangle$B InP substrates as a function of solution supercooling (ΔT_s) (after Hsieh *et al* [96]).

It can be seen that although several factors in LPE growth can affect the composition of the growing layer, they are all parameters which behave in a predictable way and are all well within the scope of the crystal grower to control.

Chapter 5

Device Applications of LPE-grown III–V Materials

5.1 Introduction

A discussion of the device applications of LPE-grown III–V materials can conveniently be subdivided into optoelectronic and microwave areas, of which the former is by far the more important. These areas can be subdivided for discussion as follows:

(i) optoelectronic devices	light emitting diodes
	lasers
	photodetectors
	optoelectronic integrated circuits
	solar cells
(ii) microwave devices	Gunn devices
	field effect transistors (FETs)
	transferred electron devices
	IMPATT diodes
	heterojunction bipolar transistors.

5.2 Optoelectronic Devices

5.2.1 Light emitting diodes (LEDs)
One of the earliest applications of LPE was to the growth of material for LEDs, initially in the visible part of the spectrum for display applications. The device utilises the electroluminescence produced when minority carriers injected across a p–n junction recombine, emitting light of a wavelength which normally (but not always) corresponds closely to the bandgap energy as shown

141

in figure 5.1, which also shows how radiative recombination associated with a deep level can be in competition with the band-to-band recombination. There is also competition with a variety of non-radiative processes such as Auger recombination, where the energy of the electron–hole recombination is passed to a third particle (electron or hole) as kinetic energy, or recombination at deep doubly ionised levels such as GaP:Cl where the doubly ionised Cl^{2+} can efficiently trap a hole.

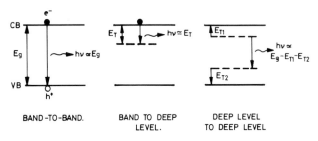

Figure 5.1 Radiative processes in light emitting diodes.

The early devices were based on GaP, with the p–n junction being produced either by p-type diffusion into an n-type layer or by double (two-layer) epitaxy, the emission wavelength being determined by the dopant species as described later. In the case of vapour-phase-grown LEDs, the alloy $GaAs_{1-y}P_y$ was used with y being varied to obtain the various colours: red ($y \sim 0.4$), yellow ($y \sim 0.86$) or green ($y = 1$). The importance of the colour of the LED emission can be appreciated from figure 5.2 which shows the relative eye sensitivity as a function of wavelength, the peak sensitivity occurring in the yellow–green part of the spectrum. Thus, given a red and green LED of equal external quantum efficiency, the green one would appear significantly brighter to the eyes.

The main problem with the GaP devices was the indirect bandgap with the accompanying low luminescence efficiencies. This handicap was partly alleviated by the use of the isoelectronic impurity nitrogen (N) in the GaP green LEDs. The excitonic recombination which occurs at the GaP:N centre has a high quantum efficiency due to the fast decay characteristic of the short-range binding potential associated with the centre. At very high N concentrations ($> 10^{20}$ cm^{-3}), the emission shifts to the yellow region of the spectrum. However, it was not possible

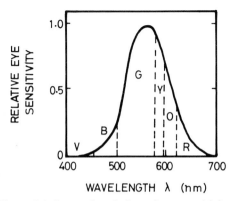

Figure 5.2 Spectral variation of eye sensitivity.

to achieve such levels of N in LPE growth, mainly due to the limited solubility of the NH_3 dopant source in the Ga solution, whereas VPE growth could achieve these high concentrations. In the case of the red LEDs, the p-layer of the device was doped with Zn and O, and the emission was due to the efficient radiative recombination of an exciton bound to a nearest-neighbour Zn–O complex. The devices normally consisted of simple one- or two-layer structures with diffused or grown junctions respectively as shown in figures 5.3(a) and (b). The p-layer is normally on top as there is less absorption of the near-bandgap light emitted by the active layer. These structures were normally grown with a vertical dipping system at 800 to 1050 °C. At such high temperatures, background contamination levels were normally high ($\sim 10^{17}$ cm^{-3}) and, together with problems of dopant loss from the solutions, cross-contamination and the difficulty in preventing loss of P from the layers, meant that the reproducibility of the devices was generally poor. This lack of reproducibility, the difficulty of scaling the process up to production, the inefficient use of Ga melts and the greater versatility of the VPE GaAs$_{1-y}$P$_y$ technology gradually led to the dwindling of interest in LPE GaP.

More recently, interest in red LEDs has concentrated on Ga$_{1-x}$Al$_x$As devices with active layers of $x \sim 0.35$ [264–267]. This material is direct gap and hence has a high luminescence efficiency. Also, the (GaAl)As/GaAs system offers the possibility of lattice-matched heterostructures. For example, Tsai and Wu [267] have described a high-brightness red LED using the

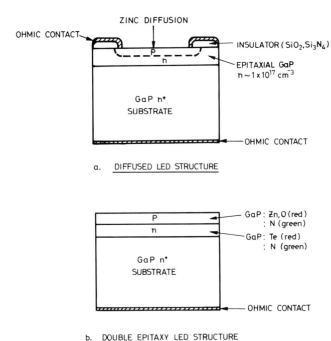

Figure 5.3 Light emitting diode structures fabricated by (*a*) diffusion and (*b*) double epitaxial growth.

structure shown in figure 5.4, with the $x = 0.7$ layers providing (i) low recombination velocity interfaces, (ii) charge carrier confinement, (iii) increased injection efficiency of minority carriers into the active layer and (iv) a transparent top layer which avoids the problem of re-absorption of the emitted radiation and hence increases the external quantum efficiency.

The early interest in the use of visible LEDs for display applications, which was very strong in the late 1960s and early 1970s and which provided the impetus for early work on LPE, has decreased due to the success of the recent developments in the competitive technologies of liquid crystal, plasma CRT and electroluminescent displays. For domestic products such a wristwatches and calculators where low power consumption is critical the LED has lost out to liquid crystal displays. In larger area displays, e.g. for home or office computers, the CRT has won out for its multicolour capability and its high number of pixels.

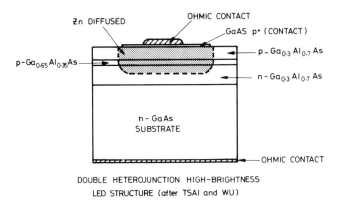

DOUBLE HETEROJUNCTION HIGH-BRIGHTNESS
LED STRUCTURE (after TSAI and WU)

Figure 5.4 High brightness red light-emitting diode structure (after Tsai and Wu [267]).

Also, electroluminescent (EL) panels based on ZnS phosphor powders or thin films have made an impact with their cheapness and high brightness.

High-radiance GaInAsP LEDs operating in the 1.3–1.6 μm region have potential application in high data-rate fibre-optic communication systems [268]. The greater linewidth of the LED source compared to the laser devices becomes less important at these longer wavelengths than for GaAs/(GaAl)As devices working at 0.9 μm due to the lower fibre dispersion in this region. The device structure can be simple consisting of n-InP substrate/n-InP/p-GaInAsP active/p-InP. By careful control of the active layer doping, modulation bandwidths of ~ 200 MHz can be achieved. The active layer thickness is not so critical as in double-heterostructure lasers, which has a beneficial effect on the device yield. Although infrared semiconductor lasers with their high output powers have become accepted as the principal sources for optical communication systems, the LED has advantages such as lower cost and higher stability.

Table 5.1 lists the various types of LED available, their emission wavelengths and typical light output measured in lumen A^{-1}. Only those marked with an asterisk have established a large commercial market. One of the major reasons for this in the case of LPE-grown devices has been the difficulty of reproducing the excellent research results in a production environment together with lower throughput of material. Thus, although LPE has consistently produced material with low concentrations of lattice defects, the poor reproducibility of background purity of GaP has been the major stumbling block in producing consistently good minority carrier lifetimes and high luminescence efficiencies.

Table 5.1 LED devices and performances.

LED material	Growth technology	Emission wavelength (μm)	Luminous output (lumen A^{-1})
GaP:N	LPE	0.57	0.3
GaP:Zn,O	LPE	0.69	0.45
GaAs$_{0.6}$P$_{0.4}$*	VPE	0.65	0.13
GaAs$_{0.35}$P$_{0.65}$:N*	VPE	0.63	0.4
GaAs$_{0.14}$P$_{0.86}$:N	VPE	0.59	0.4
GaAs:Si* (amphoteric doping)	LPE	0.93	—
GaAs/(GaAl)As (active)	LPE	0.88	—
Ga$_{0.6}$Al$_{0.35}$As/Ga$_{0.3}$Al$_{0.7}$As (active)	LPE	0.66	up to 2.6

* commercially important.

5.2.2 Semiconductor lasers

The development of the semiconductor laser, from the GaAs p–n homojunction device of 1962 [269] through single and double-heterostructure GaAs/(GaAl)As devices [270] to the double-heterostructure InP/GaInAsP laser of the 1980s [271], has gone hand-in-hand with the development of the LPE technique. In fact, the increasing performance demands of this device over the years have provided major stimuli to the research of LPE in many areas.

The semiconductor laser works, by analogy with its solid state or gas counterparts, by the stimulated emission of coherent radiation brought about by population inversion. This inversion is produced in the simplest case by passing a current at forward bias across a p–n junction, where both the n- and p-regions are degenerately doped (figure 5.5). A Fabry–Perot cavity is produced by a pair of cleavage planes as shown in figure 5.6.

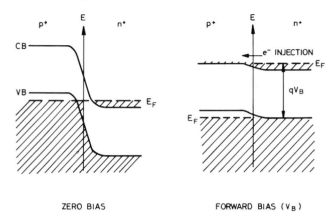

Figure 5.5 Band structure showing operation of semiconductor p–n junction laser.

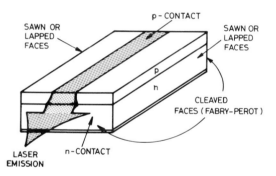

Figure 5.6 Semiconductor laser chip with cleaved-face Fabry–Perot cavity.

Initially, spontaneous emission will be obtained as in the case of an LED device. However, above a certain threshold current density, J_{th}, stimulated emission is obtained near the junction in the region of population inversion. For the early GaAs homojunction devices, the values of J_{th} were large ($10–100 \, \text{kA cm}^{-2}$) since the carrier injection was not confined. The developments in laser structures since then have had several aims:

(i) *Reduction in J_{th}.* This can be achieved by producing carrier and optical confinement in the active layer by cladding it with layers with higher bandgap and lower refractive index. The two main materials systems used have been GaAs (active)/ $Ga_{1-x}Al_xAs$ $(x \sim 0.3)$ (cladding) and GaInAsP (active)/InP (cladding). The early planar double-heterostructure (DH) laser (figure 5.7(a)) achieved confinement in one transverse direction perpendicular to the interface plane while later versions confined the lasing region by use of defined contact regions (figure 5.7(b)). More recent devices such as the double-channel planar buried-heterostructure (DCPBH) laser (figure 5.7(c)) or the inverted-rib waveguide laser (figure 5.7(d)) have confinement in both transverse directions. These complex structures often involve two steps of LPE where the first is the growth of a planar double heterostructure, and the second involves regrowth in channels etched in the original structure. This second stage exploits the ability of LPE to infill channels or grooves and the selective growth of materials, as described in §4.8. For example, let us examine in detail the growth of the DCPBH structure in figure 5.7(c) [272]. The stages in the growth of this are shown in figure 5.8. Firstly a three-layer structure is grown on a (100) InP substrate, consisting of (1) n-InP (buffer), (2) p-GaInAsP (active) and (3) p-InP (cladding). A masking layer of photoresist is then deposited and a pair of channels formed by opening long narrow windows and then etching to form round-bottomed 5–7 μm wide channels with the upper part bounded by $\langle 111 \rangle$A planes. After careful surface cleaning, the second LPE growth cycle is carried out. This consists of p-InP and n-InP (blocking layers), p-InP (confining layer) and p$^+$-GaInAsP (contact layer). The asymmetrical growth of the blocking layers whereby they do not grow on the mesa top is cleverly engineered by control of the solution supersaturation and the fact that solute depletion occurs in the solution close to the mesa-top due to growth occurring in the channels on either side of it. It can thus be seen that the sensitivity of LPE growth rate to surface curvature, crystallographic orientations and solution supersaturation can be exploited to produce complex structures which are difficult to grow by other techniques.

Reductions in J_{th} can also be achieved by reducing the active layer thickness, d, and by increasing the quantum efficiency, η. Both of these have required improvements in the LPE growth process in terms of growth rate control and defect reduction. Typical J_{th} values in modern devices are ~ 1 kA cm^{-2}.

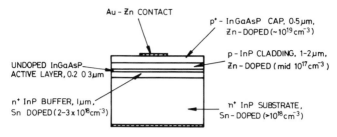

(a)   PLANAR DOUBLE - HETEROSTRUCTURE

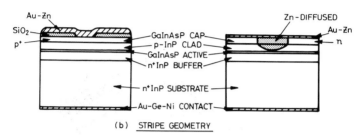

(b)   STRIPE GEOMETRY

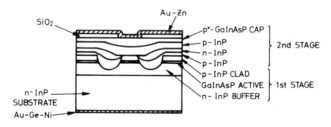

(c) DOUBLE CHANNEL PLANAR BURIED HETEROSTRUCTURE

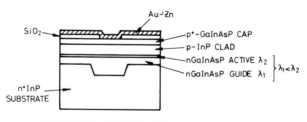

(d) INVERTED - RIB WAVEGUIDE

Figure 5.7 Various double-heterostructure laser devices of increasing complexity and degrees of confinement namely (*a*) planar confinement only, (*b*) planar plus one lateral degree of confinement and (*c*), (*d*) planar and two lateral degrees of confinement.

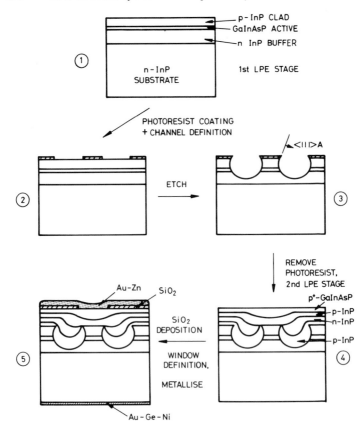

Figure 5.8 Stages in the production of a double-channel planar buried heterostructure laser using two-stage liquid-phase epitaxy (after Logan *et al* [272]).

(ii) *Increasing lifetime.* The operating lifetime of lasers is linked to the presence of dislocations and other defects in the active region of the laser. It has been shown that the degradation of both GaAs/GaAlAs [273,274] and InP/GaInAsP [275] DH lasers is directly related to the presence of dislocation networks. Dislocations in the epitaxial layers can be (a) caused by lattice mismatch between layer and substrate, (b) propagated from the substrate or (c) introduced during device fabrication. The elimination of misfit dislocations in the case of GaInAsP/ InP structures is dependent on the control of composition, which is generally very good in LPE growth as described earlier in §§4.5.1 and 4.9. Problems can however be caused by solution carry-over caused by poor wipe-off, leading to the formation of holes and regions of non-uniform composition. This has been shown to be detrimental to device lifetime [276]. Much work has been carried out on the reduction of dislocation densities in GaAs [277] and InP [278] substrate materials by heavy doping with In and S respectively and by general improvements in the technology of the liquid-encapsulated Czochralski process.

(iii) *Control of laser mode structure.* The control of the longitudinal and transverse modes of the laser is extremely important in order to (a) obtain efficient coupling of the laser output into the monomode fibres used in modern high bit-rate, long haul fibre-optic communications systems, (b) minimise dispersion effects in these fibres, (c) provide the stable mode behaviour required in several applications such as laser printing or optical disc recording/playback systems and (d) provide wavelength stability.

Several structures have been developed to achieve this mode stabilisation. One is the inverted-rib waveguide structure [259,279] (figure 5.7(d)) where the laser structure is grown on an infilled channel etched in the substrate. In this structure, it is the increase in thickness of the guide layer at the channel position which leads to an increase in the effective dielectric constant and hence to waveguiding. By careful control of the channel depth and width and the composition of the guide layer, the zero-order lateral mode is selectively guided while the higher order modes are lost. This structure has the advantage of being simple to grow with a single stage of liquid-phase epitaxy. Laser outputs of up to 25 mW per facet have been achieved in 1.3 μm GaInAsP/InP lasers whilst maintaining single lateral-mode operation. Similar channel structures have also been used in GaAlAs/GaAs lasers [280,281].

Another structure which has received much attention [282–284] is the distributed-feedback (DFB) laser, one variant of which is shown in figure 5.9. This is grown by one-step LPE onto an n-type InP substrate with a corrugated surface, the layers consisting of n-GaInAsP (active), p-InP (clad layer) and p$^+$-GaInAsP (contact layer). The corrugations have a period of

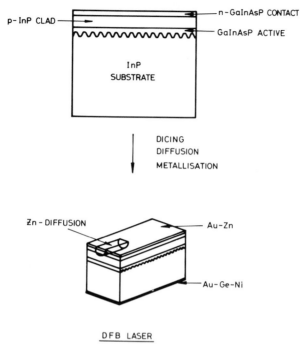

Figure 5.9 Layer structure and device structure of a distributed feedback laser.

~ 0.2–0.3 μm depending on the lasing wavelength and are normally produced by chemical etching through a photoresist mask defined by holographic photolithography. However, dry etching techniques such as ion-beam milling [285] or reactive ion-beam etching [286] have also been used. The major problem in the LPE growth of this structure has been the tendency for the corrugations to smooth out during the period at high temperature prior to commencement of LPE growth due, it is believed, to mass transport through the vapour phase [287], while Nelson *et al* [283] have shown that this deformation process is related to the presence of PH_3 in the ambient gas. The use of a GaAs cover slice instead of the usual InP to protect the substrate surface prior to growth [288] has been shown to reduce the deformation of the corrugations. The chief attribute of the DFB laser is the longitudinal mode control that it offers. This mode control is difficult to achieve in the normal Fabry–Perot cavity used in semiconductor lasers where several longitudinal modes are excited. This results in a set of laser lines spaced by ~ 1 nm. By incorporating a periodic variation in refractive index in the active region of the laser by means of the corrugations, Bragg diffraction can occur which provides spectral selectivity and feedback. The Bragg back reflection condition is

$$P = m\lambda_L/2n \qquad m = 1, 2, 3 \ldots \qquad (5.1)$$

where P is the period, m is the grating order, λ_L the laser wavelength and n the refractive index of the active layer material. For $\lambda = 1.3 \, \mu\mathrm{m}$ and a value of $n \sim 3$, then P will be $\sim 0.4 \, \mu\mathrm{m}$ for a second-order $(m = 2)$ grating. Only the wavelength corresponding to the Bragg condition is selected and hence single longitudinal-mode operation occurs.

Although we have concentrated in this section on GaAs/(GaAl)As and GaInAsP/InP lasers, there have been some publications on other materials systems. Nahory *et al* have reported on $GaAs_{1-x}Sb_x/Al_yGa_{1-y}As_{1-x}Sb_x$ DH lasers emitting at $\lambda \sim 1.0 \, \mu\mathrm{m}$ in CW mode at room temperature with low J_{th} $(2.1 \, \mathrm{kA \, cm^{-2}})$ [289], while Dolginov *et al* [290] have produced DH lasers with $Al_xGa_{1-x}As_ySb_{1-y}$ $(x < 0.1)$ active layers and wider gap $Al_{x'}Ga_{1-x'}As_{y'}Sb_{1-y'}$ confining layers grown on GaSb substrates by LPE. These had J_{th} as low as $1.6 \, \mathrm{kA \, cm^{-2}}$ at room temperature and emission wavelengths in the range 1.67–1.78 $\mu\mathrm{m}$. Room temperature DH lasers based on $Ga_{1-x}In_xAs_ySb_{1-y}$ active layers with AlGaAsSb confining layers have also been reported operating in pulsed mode at $\lambda = 2.2 \, \mu\mathrm{m}$ with $J_{th} \sim 7 \, \mathrm{kA \, cm^{-2}}$ by Caneau *et al* [291].

From the discussion above, it can be seen that the increasing sophistication of semiconductor laser structures has gone hand-in-hand with new developments in LPE technology to the extent that, at the present time, lasers represent the most important production application of LPE.

5.2.3 *Quantum-well lasers*
The ability of current epitaxial growth technologies to produce very thin layers (< 50 nm) wherein quantum size effects are observable, has led to a whole new area of research on the basic physics and device applications of so-called 'quantum wells' and

'superlattices' [292–294]. A 'superlattice' consists of alternating thin layers of two materials (normally lattice matched) such that the thickness of the layers is less than the carrier mean free path. The superlattice has a periodic potential in a direction normal to the layer structure, as shown diagrammatically in figure 5.10 for a GaAs–(GaAl)As structure. The so-called 'quantum well' produced in the narrower-bandgap material confines the electron and hole motions and produces a series of confined particle levels, E_n, with

$$E_n = \frac{h^2 n^2}{8mL^2} \qquad n = 1, 2 \ldots \qquad (5.2)$$

where h is Planck's constant, m is the electron mass and L is the well width. This produces a change in the density of states as illustrated in figure 5.11. For a free electron able to move in three dimensions in a bulk semiconductor with parabolic energy bands, the density of states $D(E)$ is proportional to $E^{1/2}$ as shown in the broken curve. In the case of an electron confined in a quantum well to move in only two dimensions, the density of states has a step structure as shown in the full line in figure 5.11, where E_1, E_2 etc are the energies of the confined particle states in the quantum well.

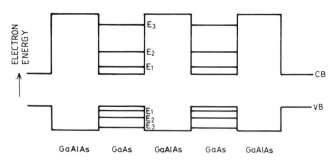

Figure 5.10 Band structure of a multiple-quantum-well device showing the origin of the confined energy levels in the wells.

The principal advantages of these quantum size effects for semiconductor lasers are threefold. Firstly, because of the sharp steps in the density of states at the energies E_1 etc, the optical

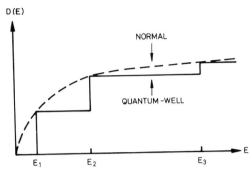

Figure 5.11 Density of states $D(E)$ as a function of electron energy, E, for a quantum-well structure compared to a normal homogeneous semiconductor.

gain in the laser is confined to a smaller range of energies giving narrower spectral linewidths. Secondly, since the energies E_1, E_2 etc are dependent on the inverse square of the quantum-well width L (equation (5.2)) the lasing emission wavelength can be tuned over quite a wide range [294]. Thirdly, radiative transition rates are enhanced in the quantum well due to the close confinement of the electrons and holes leading to reduced carrier lifetimes. This allows the laser to be modulated extremely rapidly which is of great advantage in fibre-optic communications systems, and also leads to significant reductions in the threshold current density compared with conventional double-heterostructure laser devices [295].

It has been found that single-quantum-well laser structures do not perform as well as multiple-quantum-well devices, particularly when the quantum-well thickness L_z is < 10 nm. This can be due to band-filling which produces a spectral spread of the recombination and hence reduces gain and increases the threshold current density. Another problem is the inability of excess carriers produced in the confining layers to scatter into the lower confined-particle states of the quantum well when $L_z \sim l_p$, where l_p is the electron scattering path length [296]. This problem can be overcome by using several quantum wells coupled by thin intervening barrier layers, thus making the effective active region thickness $\gg l_p$.

In comparison with MBE and MOVPE, the LPE technique has

not been widely used in the growth of GaAs–(GaAl)As quantum-well structures due to the problems of layer thickness control. However, due to the difficulties of growing GaInAsP by MBE and the relatively immature MOVPE technology for quaternary growth, much research has been carried out on the LPE growth of GaInAsP/InP quantum-well lasers [61,237,297–300]. The growth of multiple-quantum-well structures, which are needed for the reasons discussed above, involves the precise control of the thickness of the confining layers as well as the quantum-well layers. A typical structure is illustrated in figure 5.12 from the work of Sasai *et al* [61]. The active layer consists of five well layers of 20 nm thick undoped InGaAsP ($\lambda = 1.3 \,\mu$m) with four intervening barrier layers of 40–60 nm thick GaInAsP ($\lambda = 1.1 \,\mu$m). Other workers have used InP barrier layers [298,300] instead of higher-bandgap GaInAsP. Growing these structures has required the advanced LPE techniques discussed earlier in §4.6 using low growth temperatures, controlled supersaturation and computer-controlled slider movements. When fabricated into devices, such structures have produced GaInAsP ($\lambda \sim 1.3 \,\mu$m) lasers with low threshold currents [61], reduced temperature dependence of threshold current density [61,299] with $T_0 > 150$ K compared to normal values of 60–70 K (where T_0 is the coefficient in the expression $J_{th} = A \exp(T/T_0)$) and reduced frequency chirp [299] (~ 0.5 Å compared to 1–1.5 Å for laser structures with a conventional active layer).

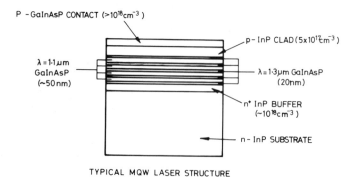

TYPICAL MQW LASER STRUCTURE

Figure 5.12 Multiple-quantum-well (MQW) laser structure in the GaInAsP/InP system (after Sasai *et al* [61]).

It is too early in this field to assess the long-term role of LPE since no figures on the yields of working devices have yet been published and progress on the growth of quaternary materials by metal-organic vapour-phase epitaxy may make this technique more attractive in the long run because of its better control of growth rate and fast switching capability.

5.2.4 Photodetectors

A photodetector device functions by absorbing incoming photons which create mobile charge carriers in the material. There are several categories of photodetector which differ in the way in which the charge is detected. The principal categories are as follows:

(i) *Photodiodes.* The charge is collected by a p–n junction. If the diode is operated at high reverse bias, the photo-induced charge carriers are accelerated and can release other carriers producing a current gain. This device is called an avalanche photodiode (APD).

(ii) *Photocathodes.* The minority carrier electrons produced by the photons diffuse to the material surface, which if specially treated to produce negative electron affinity (NEA) can allow a significant proportion of the electrons to be emitted into the vacuum.

(iii) *Phototransistors.* The carriers produce a change in the conductance of the transistor.

(iv) *Photoconductors.* The released charge changes the electrical conductivity of the material which is then detected.

In the following sections, we shall discuss each of these. However, as far as epitaxial III–V materials are concerned, the principal applications have been in the photodiode and photocathode categories. The areas of application of photodetectors are many and varied, being mainly in the infrared region of the spectrum. The principal wavelengths of interest in military and civilian applications are listed in table 5.2. The great advantages that the III–V materials offer over Si and Ge for photodetector applications are (i) the ability to vary the bandgap of the material to suit a particular application by the use of ternary or quaternary alloys in order to reduce thermally generated dark currents, (ii) the fact that the materials are generally direct gap

Table 5.2 Wavelengths of interest for semiconductor photodetectors and the associated applications.

Wavelength (μm)	Applications
< 0.9	Detectors for visible/near-IR lasers in video-disk, compact disk systems
0.9	Fibre-optic communications systems based on GaAs/GaAlAs lasers
0.9–1.0	Photocathodes based on GaAs:Cs,O for night vision using ambient night-sky radiation
1.06	Use with Nd:YAG lasers in laser range-finding
1.3–1.6	Fibre-optic communications using low-loss, low-dispersion silica fibres for long haul, high bit rate system
2.06	Eye-safe laser range-finder in conjunction with Ho:$\alpha\beta$ YLF laser
2–3	Fibre-optic communications using fluouride glass fibres
3–5	Thermal imaging and missile guidance
8–13	Thermal imaging and missile guidance

with high absorption coefficients and (iii) their high mobilities offer the possibility of high-speed detection. In the red and near-infrared region up to 1.1 μm, the Si APD device is currently in a dominant position mainly because of the very low excess noise in these detectors. For wavelengths above 1.1 μm, the III–V materials have come into their own and the major efforts recently have been on the 1.3 μm and 1.55 μm wavelengths for fibre-optic communication purposes. Longer wavelength III–V detectors in the 2–10 μm range have not until recently been widely studied apart from InSb, which was widely researched in the late 1950s mainly in the form of bulk-grown single crystals. At the time, it was the smallest bandgap semiconductor material known with $E_g = 0.17$ eV, $\lambda_{cut-off} \sim 7$ μm at room temperature. The smallest bandgap available in the III–V alloy systems is in the InAs–InSb ternary which shows a pronouced negative bowing with a minimum of 0.1 eV at a composition of InAsSb, corresponding to $\lambda_{cut-off} \sim 9.0$ μm at 77 K. However, this composition is not lattice matched to a binary substrate and compositional grading layers are required. For wavelengths around

10 μm and above, other materials systems such as the II–VI alloy system $Cd_xHg_{1-x}Te$ (CMT) or the IV–VI alloy $Pb_{1-x}Sn_xTe$ (LTT) have been predominant. Thus it is in the short wavelength region of the IR from 1–5 μm between the operating wavelengths of Si and CMT/LTT where the III–V compounds and alloys have been most used in photodetector applications. The need for materials to cover this range has been a major stimulus to the LPE growth of III–V alloys.

Photodiodes

The key parameters of a photodiode for use as a receiver in optical communications systems in conjunction with typically GaAs MESFET amplifier are (i) high quantum efficiency, (ii) fast response time, (iii) low noise, (iv) low dark current and (v) uniform gain in the case of avalanche photodiodes.

There are several types of photodiode device structure which exploit the potential of epitaxial growth and these are illustrated in figure 5.13(a) for the case of $Ga_{0.47}In_{0.53}As$. There is the simple p–n or p–i–n homojunction detector which is grown by multiple LPE using Zn for the p-dopant and Sn or Te for n-type doping [56,301]. The layers are all 2–4 μm thick which is a favourable value for LPE growth. This structure has the advantage of exact lattice matching at the p–n junction and minimised interface states leading to low reverse leakage currents.

The first layer to be grown is normally a buffer layer of either InP or $Ga_{0.47}In_{0.53}As$ to ensure that any defects associated with the substrate interface are remote from the depletion region. The second (ternary) layer is undoped and needs to be n-type with a carrier concentration of $\sim 10^{15}$ cm^{-3} or less, while the final (ternary) layer is p$^+$ at $\sim 10^{18}$ cm^{-3}. The achievement of low carrier concentration in the n$^-$ layer requires careful baking of the growth solution. The p–n junction in such devices can also be produced by diffusion of Zn into the n$^-$ layer [302] or by ion implantation [303]. The detectors are normally fabricated by etching mesas and then making electrical contacts to the p-region (top of the mesa) and the n-region (on the back of the substrate). Illumination can be either from the front or back-side, with the latter having the advantages of reduced absorption loss and surface recombination [302]. Similar homojunction device structures have been used for other LPE-grown III–V

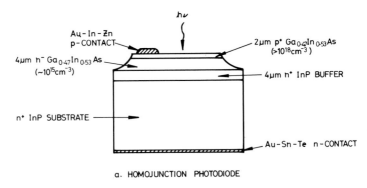

a. HOMOJUNCTION PHOTODIODE

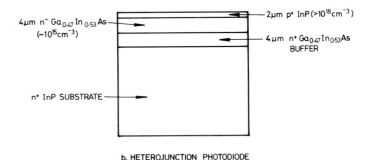

b. HETEROJUNCTION PHOTODIODE

Figure 5.13 Various epitaxial photodiode structures: (*a*) epitaxial $Ga_{0.47}In_{0.53}As/InP$ homojunction structure; (*b*) heterojunction $Ga_{0.47}In_{0.53}As/InP$ photodiode.

materials such as GaInAsSb ($\lambda = 2.3$ μm) [304], GaAsSb ($\lambda = 0.90$–0.98 μm) [154], InAsSb ($\lambda \sim 4$ μm) [156,305] and GaInAsP ($\lambda \sim 1.4$ μm) [306].

Another generic device type is the heterojunction photodiode an example of which is shown in figure 5.13(*b*) for the case of (GaIn)As/InP. This device consists of an n^+-InP substrate on which is grown an n^+-$Ga_{0.47}In_{0.53}As$ buffer layer as before, an n^--$Ga_{0.47}In_{0.53}As$ active layer and a p^+-InP window layer. There are two main advantages of this type of structure. Firstly, the lattice-matched InP layer reduces the effects of surface recombination which would normally lower the quantum efficiency of the photodiode at photon energies above the bandgap. Secondly, the heterojunction device should produce an enhanced response time since all the photons are absorbed in the depletion region near the p^+-InP/n^--Ga$_{0.47}$In$_{0.53}$As interface and move rapidly across the depletion region, whereas minority carriers produced in the 'bulk' semiconductor diffuse more slowly to the junction. The heterojunction needs to be very precisely lattice matched to avoid introducing mismatch dislocations which would act as non-radiative recombination centres. Moon [154] has shown a clear correlation between the performance degradation of GaAsSb photodiodes and increasing threading dislocation density. The good control of alloy composition in LPE growth is a great advantage in this respect, and because of the versatility of LPE in growing a wide range of lattice-matched ternaries and quaternaries, it has been widely used for growing these device structures in various materials systems such as GaInAsP/InP [307,308,309], InAsSb/GaSb [310], GaInAs/InP [56], GaAlSb/GaSb [311,312] and GaAlAsSb/GaSb [313].

The third generic type of photodiode is the avalanche photodiode (APD) in which the photo-injected carriers are accelerated in the electric field and can undergo collisions to create more carriers and so on to produce an avalanche effect which amplifies the original signal. The avalanche process is obtained by operating devices similar in structure to those discussed above in large reverse bias. The avalanche gain is characterised by the ionisation rates α, β for electrons and holes which are dependent on electric field strength and orientation. A major problem with APDS is the excess noise produced in the avalanche multiplication process and this is at its worst when $\alpha/\beta \simeq 1$. It is therefore most favourable for excess noise reduction that $\alpha/\beta \gg 1$ or $\alpha/\beta \ll 1$, and that the reverse leakage current be minimised. However, the values of α, β are fixed for a particular material and so various devices have been designed to ensure that at the large fields required to achieve avalanche gain (10^5–10^6 V cm^{-1}) the tunnelling currents are minimised. One way of achieving this is with separate absorption and multiplication (SAM) which has been used in the InGaAs/InP system [314]. A typical structure is shown in figure 5.14 together with the band structure diagram. Here, the optical absorption occurs in the InGaAs layer where the field is kept low to reduce tunnelling, while the multiplication process occurs in the InP where there is a high field. Such SAM devices produce an improvement in performance in comparison with normal p–i–n devices. Problems with slow response in such devices caused by trapping of holes at the discontinuity in the valence band shown in figure 5.14 have been overcome by putting an n^--GaInAsP layer between the n^--GaInAs (active) and n^--InP (multiplication) layers [315]. Structures have also been devised which enhance the value of the α/β ratio using superlattices, graded-gaps and staircase structures [316] though these more complex structures require MBE or MOVPE growth rather than LPE. In order to sustain the high electric fields required for avalanching without reaching breakdown, the active layer doping must be uniform and fairly low since the reverse breakdown voltage (V_B) increases as the carrier density is reduced. The layer thickness and composition need to be uniform. Also, the density of dislocations must be low in order to avoid the formation of hot spots and microplasmas which reduce the avalanche gain.

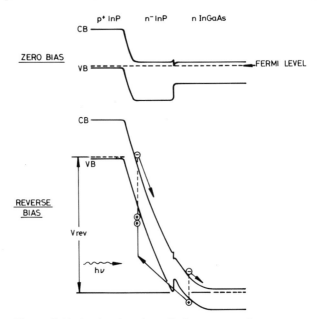

Figure 5.14 Avalanche photodiode structure incorporating separate absorption and multiplication (SAM) in the (GaIn)As/InP system (after Susa *et al* [314]).

The use of heterojunction APD structures with a wide bandgap window layer enables high quantum efficiencies and high response speeds to be obtained, by ensuring that nearly all the radiation is absorbed in the depletion region where the carriers travel near the saturated drift velocity. For example, Tomasetta *et al* [317] have grown a GaAlAsSb/GaAlSb heterojunction APD structure working in the $\lambda = 1.0$–$1.4\,\mu\text{m}$ region with an internal quantum efficiency of 90% and a risetime of 60 ps. It is vital that the hetero-interface is highly perfect with a low density of dislocations which could act as recombination sites for the minority carriers.

In summary, it can be seen that photodiode structures place several demands on the materials grower:

(i) high purity in the active layer,
(ii) long minority carrier diffusion lengths,
(iii) low interface recombination velocities implying low mismatch at heterojunctions and
(iv) uniformity of active layer thickness, composition, and carrier concentration.

The LPE process can meet these requirements well, particularly as the layer thicknesses are typically 1–5 μm, which is in the optimum range for LPE growth. However, the future novel detector structures requiring bandgap engineering will not be easily produced by LPE and will be better grown by MOVPE or MBE.

Photoconductors

Apart from InSb, very little work has been reported on the use of III–V materials for photoconductive detectors. Even less has been reported on LPE-grown III–V materials. Although photoconductive detectors have certain advantages such as internal gain and simplicity of structure, they do place severe demands on the material quality. In particular, the material must have a low carrier concentration, high mobility and long minority carrier lifetimes. Recently, the applicability of photoconductive detectors to fibre-optic communication systems has been investigated either using GaAs [318,319] or $\text{In}_{0.53}\text{Ga}_{0.47}\text{As}$ [320,322], the hope being to produce low-noise, very high-speed detectors. By using LPE-grown InGaAs on InP substrates, Degani *et al* [322] were able to produce extremely fast detectors with rise and fall times of 45 and 75 ps respectively at 5 V bias. Chen and Chi [323] were able to obtain a reduction in dark current of a factor of over 200 by the use of Fe doping the active LPE-grown layer compared to undoped layers of $Ga_{0.47}In_{0.53}As$. This led to a reduction in noise current of 3.7 dB. The effect of the Fe doping was to reduce the carrier concentration by ~ 100 to $\sim 1.8 \times 10^{13}$ cm^{-3} while producing a modest reduction in room temperature mobility from 8×10^3 to 5.4×10^3 cm^2 V^{-1} s^{-1}. Although a reduced gain was obtained, the very low noise capability could be important for low bit-rate applications.

Phototransistors
There has been some interest in the use of phototransistors as alternatives to avalanche photodiodes plus FETs in fibre-optic communication systems, the aim being to achieve high gain without the problems of excess noise encountered in APD devices. Figure 5.15 shows a heterojunction bipolar phototransistor structure grown by LPE in the GaAs/(GaAl)As system [324], which achieved current gains in the region of 40,000 to 50,000, among the highest values ever reported for such a device. This was attributed to the high electron lifetimes achieved in the p$^+$-base region.

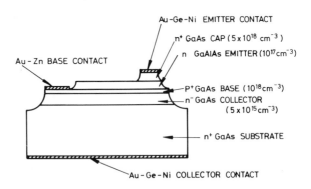

HETEROJUNCTION PHOTO-TRANSISTOR

Figure 5.15 Heterojunction bipolar phototransistor structure in the (GaAl)As/GaAs system (after Cazarre *et al* [324]).

Impressive results have also been reported for InGaAs/InP HBT devices grown by LPE, with high optical gains ($\sim 6 \times 10^3$) [325] and fast response times (~ 180 ps) [326].

Phototransistor devices have also been incorporated in more complex integrated device structures and these will be discussed in the following section.

Optoelectronic integrated circuits (OEIC)

An increasingly important aspect of the recent research in high-speed fibre-optic communications is the integration of optical components such as lasers and photodetectors with electronic components such as FETs on a single chip, first in the GaAs/(GaAl)As system [327–329] and more recently in the longer-wavelength GaInAsP/InP system [330–332]. Both these systems offer materials with high electron mobilities and efficient luminescent transitions which are required for good FET and laser diode performance, respectively.

The benefits of such integration are chiefly (i) higher operating speed due to reduced parasitic impedance, (ii) improved reliability and (iii) reduced production costs.

There are two approaches to integration, the first involving the growth of all the required layers in a vertical sequence such as that shown in figure 5.16(a) or by growing the separate device structures horizontally separated as shown in figure 5.16(b) [332]. This was grown in a two-stage LPE process whereby the laser structure was grown in the first stage followed by photolithographic definition of the laser stripe which was then buried in a second LPE process with the layers of the heterojunction bipolar transistor structure. The latter approach exploits the strengths of LPE better since there are fewer layers in each stage of the growth sequence, thus simplifying the boat design, and the technique of growth on structured substrates can be used. The LPE process imposes certain constraints on the complexity and variety of OEIC device structures that can be grown due to limitations on the number of solutions in the boat, the abruptness of doping changes achievable and the problems of melt-back of previously grown layers. In spite of these difficulties, extremely sophisticated devices have been successfully grown by LPE and have shown great promise [332].

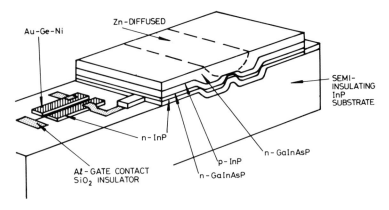

(a)   MONOLITHIC STRUCTURE BY SINGLE STEP L.P.E

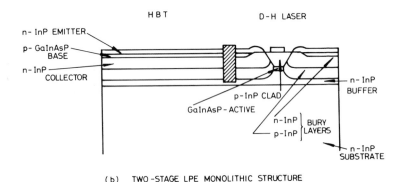

(b)   TWO-STAGE LPE MONOLITHIC STRUCTURE

Figure 5.16 Optoelectronic integrated circuit structures: (*a*) a monolithic structure grown by single-step LPE, and (*b*) horizontally separated structure grown by two-stage LPE (after Shibata *et al* [332]).

An interesting recent development is that of the laser transistor device described by Mori *et al* [333] and shown in figure 5.17. This is a buried mesa structure grown by two-step LPE using the InP/GaInAsP system. This device can operate as a double-heterostructure laser or a heterojunction bipolar transistor. The p-GaInAsP layer can act as either the active layer of the laser or the base of the HBT. By modulating the collector voltage, the laser output can be switched at potentially much higher speeds than can normally be achieved. This device could form a key part of high speed OEICs.

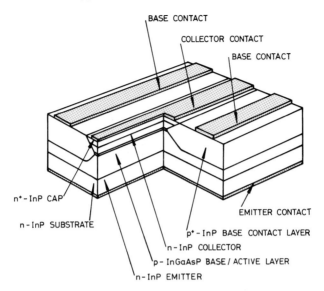

BASE CONTACT

COLLECTOR CONTACT

BASE CONTACT

n⁺-InP CAP

n-InP SUBSTRATE

EMITTER CONTACT

p⁺-InP BASE CONTACT LAYER

n-InP COLLECTOR

p-InGaAsP BASE / ACTIVE LAYER

n-InP EMITTER

Figure 5.17 Laser transistor structure in the InGaAsP/InP system grown by two-stage LPE (after Mori *et al* [333]).

Photocathodes

The work of Scheer and van Laar using caesiated GaAs [334] and that of Turnbull and Evans [335] with GaAs (Cs,O) demonstrated the potential of producing a negative electron affinity (NEA) photocathode imaging tube with single crystal III–V materials. These so-called third-generation photocathodes were expected to have several advantages over the second-generation multi-alkali photocathodes namely:

(i) extended reponse in the near IR which is of importance for military night-vision applications using the available night-sky radiation,

(ii) improved sensitivity obtained through the use of single crystal materials with long minority carrier diffusion lengths and the fact that thermalised electrons could be emitted whereas hot electrons only are emitted in the multi-alkali photocathodes, and

(iii) improved spatial resolution due to the low spread of transverse electron energies.

Although much work has been carried out on a wide variety of III–V materials, it is only the GaAs:Cs,O photocathode that has been produced commercially so far.

Figure 5.18 shows the principle of operation of the GaAs: Cs,O photocathode in the transmission mode using a schematic of the band structure. Photons enter the rear of the GaAs layer through a suitable transparent support and create minority carrier electrons in the heavily p-type GaAs layer. A certain proportion of these carriers diffuse to the negative electron-affinity surface where they are emitted into the vacuum and can be imaged on a phosphor screen. Two important parameters which increase the photocathode efficiency are (i) a high minority carrier diffusion length in the p^+-GaAs and (ii) a low recombination velocity at the interface between the GaAs and the transparent backing. While early structures used GaAs deposited on to sapphire or (GaAl)As/GaP substrates [336], the breakthrough occurred with the use of lattice-matched (GaAl)As/GaAs structures grown by LPE onto GaAs substrates as shown in figure 5.19. To obtain the photocathode active layer on a transparent (glass) substrate, the sequence of steps shown in figure 5.19 were used, the critical steps being the bonding of the semiconductor structure to glass and the removal of the substrate using selective chemical etching [337]. The LPE (GaAl)As buffer layer between the GaAs active layer and the glass ensured that there was a low density of interfacial defects and hence a low recombination velocity. Although the LPE process was able to grow such structures with excellent electrical and optical properties and hence high sensitivities, the requirement for blemish-free surfaces over the whole imaging area of the devices (~ 20 mm diameter) has proved difficult to achieve with LPE. Small pinholes in the active layer due to the incomplete solution removal from the previous (GaAl)As growth was a common feature and this led to the eventual use of cosmetically superior MOVPE material once problems of low minority carrier diffusion length had been overcome [212,338].

Solar cells
The field of solar photovoltaic energy conversion has been an extremely active one in the past decade driven by two different requirements, one being the possibility of large-scale solar power stations to provide local domestic and industrial electric power for isolated communities using large areas of very cheap (probably silicon) solar cells. The other application is the

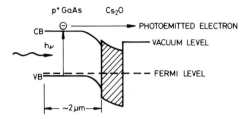

Figure 5.18 Band diagram showing principle of operation of a GaAs negative electron affinity photocathode.

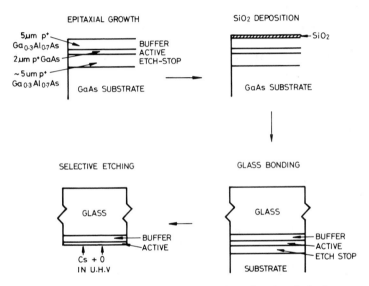

Figure 5.19 Stages in the production of a glass-bonded photocathode structure (after Antypas and Edgecumbe [337]).

provision of solar power for the satellites increasingly used for telecommunications, surveillance, resource mapping, meteorological and space laboratory applications. Here, the need for high efficiency and radiation hardness overcomes that of cheapness. It is in this area that cells based on high-quality single crystal semiconductors such as CdTe and GaAs have found their application. The GaAs–(GaAl)As solar cell has been particularly widely studied. A typical structure is shown in figure 5.20(a)

consisting of a p-GaAlAs layer on top of a p–n GaAs junction. This can be produced by growing all three layers epitaxially, by growing the p-GaAlAs and p-GaAs layers on an n-type GaAs substrate $(n = 1{-}5 \times 10^{17}\,\mathrm{cm}^{-3})$ or by growing only the p-GaAlAs layer and allowing the p-dopant, normally Zn, to diffuse into the n-GaAs substrate to produce the p–n junction. The use of the transparent GaAlAs layer produces a low recombination velocity interface with the GaAs and the step in the conduction band (figure 5.20(b)) prevents photogenerated electrons in the p-GaAs from being lost into the (GaAl)As layer. These structures are most frequently grown by liquid-phase epitaxy due to the long minority carrier diffusion lengths and low interface recombination velocities obtained. Using such LPE-grown structures, high efficiencies (up to 24%) have been obtained by several workers [339–341]. To improve the efficiencies at shorter wavelengths, absorption losses in the (GaAl)As top layer can be reduced by making it very thin ($< 50\,\mathrm{nm}$), or by producing a built-in electric field with graded bandgap (GaAl)As window layer [341]. This can be achieved by isothermal contact of a GaAs substrate with a saturated Ga–Al–As solution to produce a thin ($< 100\,\mathrm{nm}$) GaAlAs layer with highest bandgap at the surface.

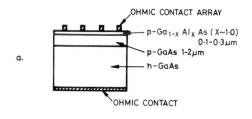

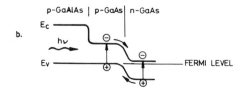

Figure 5.20 (a) Typical GaAs–(GaAl)As solar cell structure; (b) band diagram of the solar cell device.

For use under high concentration (> 200 suns) conditions, the series resistance becomes an important factor to be reduced and this can be achieved by heavily doping the p-GaAlAs window layer or by making the p-contact to the underlying p-GaAs layer by means of photolithography and selective chemical removal of parts of the window layer [340].

Further improvements in overall efficiency have been predicted by using several layers of differing bandgaps with the highest bandgap at the surface [342]. This avoids losses due to the generation of carriers at photon energies much above the bandgap which can occur when a single (GaAs) active layer is used. For solar cell structures with more than two such layers, the' use of MOVPE rather than LPE would become increasingly attractive from the point of view of yields, particularly if the advantages of large-area growth by MOVPE are considered as well.

5.3 Microwave Devices and Integrated Circuits

The discovery of the Gunn effect was a breakthrough in the application of the III–V compounds to microwave solid state devices. It was followed by the field-effect transistor (FET), the IMPATT diode and Schottky diode. The use of such semiconductor devices provided several advantages over the microwave tubes traditionally used, such as low noise, cheapness, compactness and lower operating voltages, thus opening up new applications in radar and communications, particularly in military systems. The rapid expansion of civilian satellite communications promises to provide large new markets for GaAs emitters, receivers and mixers. Another area which is likely to become increasingly important is that of GaAs integrated circuits for high-speed signal processing applications and also for fast logic (75 GHz) in future computer systems where Si devices are not fast enough. This is due to the fact that GaAs has a mobility over five times that of Si for equivalent carrier concentrations. These devices have quite stringent materials requirements, particulary in terms of high carrier mobility and good control of thickness uniformity, carrier concentration and interface abruptness. Liquid-phase epitaxy has not been the favoured way of growing the material for these devices, with the Ga–AsCl$_3$ (AsH$_3$)–H$_2$ vapour-phase epitaxy technique being the most widely used, although MOVPE is becoming more popular due to improvements in layer purity and interface control. However, some workers have reported work using LPE-grown material and this will be briefly reviewed in the following sections.

5.3.1 Gunn diodes

These are basically quite simple devices, with the earliest structures being a bar of n-type GaAs with ohmic contacts at each end. When the electric field across the bar is greater than a threshold value, typically 3–4 kV cm^{-1}, electrons are transferred from the Γ_6 minimum in the conduction band into the X_6-minimum where the room temperature mobility is 300 cm^2 V^{-1} s^{-1} compared to a value of over 7×10^3 cm^2 V^{-1} s^{-1} in the Γ_6 minimum. This reduction in mobility causes the carriers to form domains which propagate through the sample, and the transit time of the domain is related to the frequency of the device. The thickness of the GaAs is typically about 2 μm for a V-band (40–80 GHz) device and 8 μm for X-band (8–12 GHz). The need for such small device thicknesses led to the use of epitaxial growth, a typical structure being shown in figure 5.21. The n$^+$-GaAs substrate is typically mid-10^{17} cm^{-3} doped and the n$^+$ buffer and contact layers are normally in the range 10^{17}–10^{18} cm^{-3}. For the active layers, the important factors are (i) the nl product where n is the carrier concentration and l the layer thickness with typical values of $\sim 2 \times 10^{12}$ cm^{-2}, (ii) good uniformity of n and l across the device, (iii) sharp transitions between the n$^+$ and n layers and (iv) high electron mobility. The carrier concentration n is typically $\sim 10^{15}$ cm^{-3}. Many of the early devices were grown by LPE and good results were obtained with power efficiencies in the range 4–12% [343]. The LPE-grown material had good and reproducible electrical properties. The major difficulty has been in obtaining the required uniformity of nl, and particularly the layer thickness l. For this reason, material grown by the Ga–AsCl$_3$–H$_2$ VPE technique with its improved surface morphology has become favoured for such devices. The Gunn device has been replaced for many applications by the FET, which will be described in the following section.

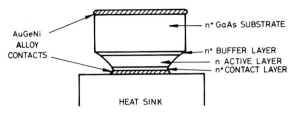

Figure 5.21 Typical epitaxial GaAs Gunn diode structure.

5.3.2 *Field-effect transistors*

In the field of silicon microelectronics, the metal–oxide–semiconductor field-effect transistor (MOSFET) has become the key component. A major reason for the success of the Si MOSFET technology in its various forms (CMOS, NMOS) has been the suitability of the native oxide, SiO_2. This situation does not occur in GaAs or the other III–V compounds as the native oxides are not stable and are of variable composition. This has led to the development of the metal–semiconductor FET (MESFET) or metal–insulator–semiconductor FET (MISFET) technologies. The basic principle of a GaAs MESFET device is shown in figure 5.22. The current in the active layer between the source and drain contacts is modulated by applying reverse bias (V_G) to the metal–Schottky gate contact. As V_G increases, the depletion region depth below the gate increases until eventually the current falls to zero at the 'pinch-off' voltage. The important device parameters are (i) the gate length which determines the maximum operating frequency of the device, typical values being in the range 0.5–2 μm (1–10 GHz), (ii) the electron mobility in the active layer must be very high right up to the interface with the buffer layer to obtain high gain and low noise (the presence of the buffer layer is necessary to avoid the poor electrical properties found at layer/substrate interfaces due to the presence of defects and contamination), (iii) the active layer thickness must be well controlled and uniform to obtain good pinch-off. The most widely used growth technique for such FET structures has been the Ga–AsCl$_3$–H$_2$ VPE process with its good purity, sharp interface control and flat surface topography. However, LPE growth has been investigated for FETs in GaAs [60,345–347], GaInAs/InP [348] and InP/InGaAsP [349,350]. In

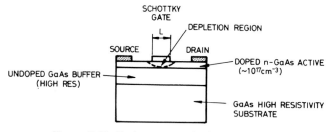

Figure 5.22 GaAs MESFET device structure.

the case of GaAs devices, the advantage of LPE-grown material was the low level of electron traps. Nanishi *et al* [345] in a comparison of high-purity LPE and VPE material for the buffer layers in FET structures found that while the LPE layers had $< 5 \times 10^{12}$ cm^{-3} of deep-level electron traps, the VPE material had $\sim 2 \times 10^{14}$ cm^{-3}. These layers used as buffer layers in FET structures gave devices with low noise figures of 2.4 dB at 10 GHz. The major difficulty has been in achieving the low carrier concentrations required in such buffer layers on a regular basis. Elaborate heat treatments of the Ga solutions are necessary with careful minimisation of the water and oxygen concentrations in the H_2 ambient gas to reduce the O and Si concentrations in the solution. Although background levels down to $n = 1 \times 10^{12}$ cm^{-3} have been reported [39] it has proved difficult to maintain these levels reproducibly. The growth of undoped GaAs buffer layers by LPE onto Cr-doped GaAs substrates was found to give excellent FET results with a gain of 10.6 dB and noise figure of 1.87 dB at 12 GHz [352]. The abruptness of the carrier concentration change between the buffer and active layers has been found to be in the range from 0.07 μm [345] to $\sim 0.2\ \mu$m [353] per order of magnitude. This is not as abrupt as achieved by MBE or MOVPE ($< 0.01\ \mu$m per order of magnitude) not because of any marked difference in interdiffusion during growth but more likely due to the slight melt-back/regrowth which occurs at the start of each new LPE layer growth. Perhaps the biggest problem in growing FET structures by LPE is to achieve reproducibility and uniformity of doping and thickness in the active layers even using the sliding boat technique. This has been highlighted in a comparative study of submicron MESFET structures produced by LPE, VPE or ion implantation techniques [347]. Although the LPE material gave comparable device performance, it was necessary to use successive anodic oxidation and stripping stages to improve the surface flatness. A similar process has been used to thin the LPE-grown active layers to the correct thickness [345,354].

In studying the growth of n-doped GaInAs lattice matched on to InP:Fe substrates for the production of JFET devices with p$^+$ diffused gates, Schmitt *et al* [348] looked at reproducibility of the GaInAs active layer properties. Their results were as follows:

Carrier concentration (average) $n = 1 \times 10^{17}$ cm^{-3} from 14 experiments

Standard deviation $\sigma(n) = 10\%$ (run-to-run)
3.2% (across wafer)

Standard deviation of experimental to theoretical thickness ratio

$$\sigma(d) = 17\% \text{ from 39 runs}$$

Standard deviation of thickness across 0.4 μm layer $\sigma(d) = 10\%$

Standard deviation of lattice mismatch $\sigma(\Delta a/a)$

$$= 1.5 \times 10^{-4} \text{ (run-to-run)}$$
$$0.82 \times 10^{-4} \text{ (across wafer).}$$

It was concluded that while the electrical and compositional homogeneity were acceptable, the variation of thickness was not. Very promising JFET results were obtained however with transconductances of 100 ms mm^{-1} and cut-off frequencies of over 30 GHz [348].

5.3.3 *InP transferred-electron devices*
Transferred-electron devices (oscillators or amplifiers) fabricated from InP are expected to have higher efficiency and lower noise than similar GaAs devices, particularly at high frequencies, exploiting the higher peak electron velocity, high low-field mobility, higher current peak-to-valley ratio and faster inter-valley scattering compared to GaAs. In addition, the higher thermal conductivity of InP favours CW operation of devices. However, the materials requirements in terms of high purity and

hence high mobility have proved a stumbling block for LPE-grown material. Difficulty in achieving net background donor concentrations of $< 10^{15}$ cm^{-3} has been chiefly attributed to the presence of Si as a major impurity in the growth solutions [200,355]. Not only is this difficult to remove by baking [355,356] but it has a high distribution coefficient [46,50]. In contrast, vapour-phase epitaxy using the In–PCl$_3$–H$_2$ process has been able to give InP material with $n < 10^{14}$ cm^{-3} and $\mu_{77} > 10^5$ cm^2 V^{-1} s^{-1} [357]. In spite of this, TE devices have been made on LPE material [358]. As in the case of GaAs FET devices mentioned in the previous section, the InP LPE layers had to be chemically thinned to obtain the correct nl product. In spite of the material having only a modest 77 K mobility of 42 000 cm^2 V^{-1} s^{-1}, good device performance as a TED was obtained with an output of 417 mW (7.6% efficiency) at 34 GHz, which compared well with current state of the art.

5.3.4 IMPATT *diodes*

The impact avalanche transit time (IMPATT) family of microwave devices operates with a p–n or Schottky (metal–semiconductor) junction (figure 5.23(a)) in reverse bias in a state of avalanche breakdown. The electrons created in the avalanche zone near the junction drift in the electric field towards the contact. A negative resistance behaviour occurs due to the transit time for charge across the drift region and the fact that the avalanche current has a phase delay with respect to the RF voltage. The frequency of operation depends on the depletion layer width and hence on the doping level. The main advantages of IMPATT diodes are the high power (up to 50 W pulsed), the high conversion efficiencies ($\sim$ 50% for the 'lo–hi–lo' structure shown in figure 5.23(b)) and operation up to high frequencies (80–90 GHz) although the available power reduces with increasing frequency. These devices are extremely demanding in their materials specification. In particular, the carrier concentration and thickness of the avalanche region need to be very well controlled and uniform. For example, in the case of 'lo–hi–lo' devices the avalanche layer thickness (n–n$^+$) needs to be $\pm$ 0.02 μm, and the total charge in the n$^+$ spike needs to be controlled in a small range ($\sim 10^{12}$ electrons cm^{-2}) for maximum

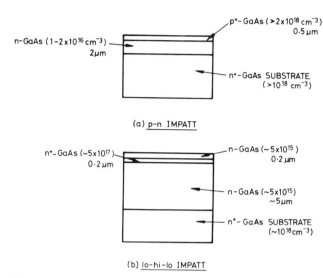

Figure 5.23 GaAs IMPATT diode structures, (*a*) p–n junction device, (*b*) lo–hi–lo device.

efficiency, hence doping level and thickness need to be simultaneously under control. Very few reports of LPE growth of these devices have appeared. Rosztoczy and co-workers have reported IMPATT results [343,351,359] on LPE-grown structures. For double-drift devices, efficiencies of 16% were obtained at 9 GHz with power output of 3 W, which is a respectable if unspectacular result. For 'lo–hi–lo' devices, efficiencies of 35% at 10 GHz with an output of up to 4 W (pulsed) were obtained, which represents a very good result. The major reason why such figures have not led to wider interest in LPE-grown structures is the difficulty in obtaining the necessary reproducibility of critical device parameters, particularly of layer thickness and doping levels in the 'hi' layers.

5.3.5 *Heterojunction bipolar transistors*

There is much interest in III–V bipolar transistor devices based on the GaAs/GaAlAs and InP/GaInAsP systems for microwave and high-speed logic at frequencies where Si bipolar transistors do not operate. These III–V devices exploit (i) the higher electron mobilities in GaAs and GaInAs hence reducing the base transit times and giving higher cut-off frequencies and (ii) the advantages of heterojunctions, in particular the use of a wide bandgap (GaAl)As emitter to improve the emitter efficiency and remove the minority carrier injection from the base into the emitter due to the potential barrier at the heterojunction. Other advantages of the III–Vs are the direct band structures which lead to short minority carrier lifetimes and hence short storage times, and the high breakdown fields leading to higher power capabilities. Disadvantages of III–V materials are (i) the difficulty of making low-resistance ohmic contacts to them and (ii) the low thermal conductivities which limit the power output of the device.

There have been several papers published describing the growth of HBTS by liquid-phase epitaxy [360–365]. A typical device structure is shown in figure 5.24(*a*). To obtain the low background carrier concentration in the n-GaAs collector level and the necessary long minority carrier diffusion lengths in the p^+-GaAs base level, careful baking of the LPE solutions was found to be necessary [362]. The current gain of the transistor, h_{FE}, is given approximately by τ_n/τ_{BT} where τ_n is the minority carrier (electron) lifetime in the p^+ base region and τ_{BT} is the base transit time. Thus the doping level and thickness of the base need to be well controlled. In the LPE growth of such structures, the reproducibility of the thickness of the base region ($\sim 0.3\ \mu$m) could prove a problem. The highly doped n^+ emitter contact layer is necessary to reduce the emitter contact resistance.

Extremely good results have been achieved with LPE-grown HBT devices. Intrinsic current gains h_{fe} have been reported in the range 1000 [362] up to 1600 [361], with emitter–collector breakdown voltages (V_{CE_0}) of up to 130 V [363]. Unit gain (f_{max}) and cut-off (f_T) frequencies were ~ 3 GHz and ~ 4 GHz respectively [362,363]. The main reason for this success was probably the lack of defects at the (GaAl)As (emitter)/GaAs (base) interface which is a feature of such heterojunctions grown by LPE.

Recently, HBTS have been fabricated from LPE-grown GaInAsP/InP [332,333,365] or InGaAs/InP [362] structures.

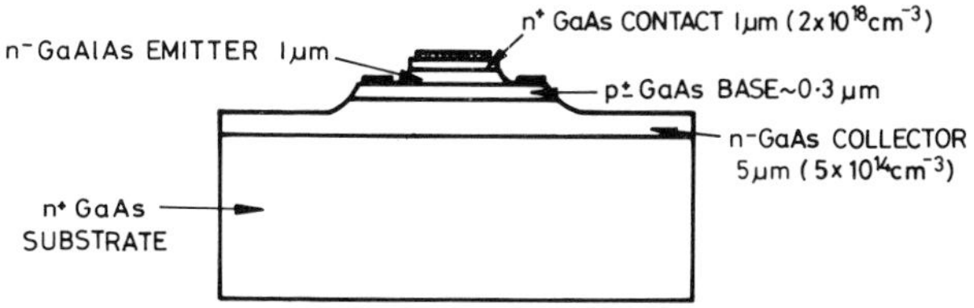

a. HETERO-JUNCTION BIPOLAR TRANSISTOR

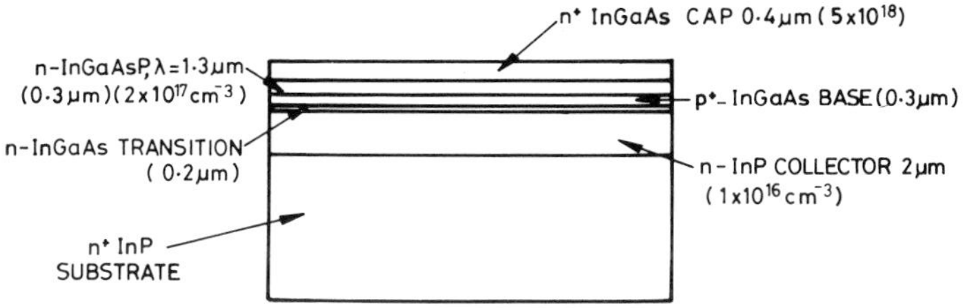

b. DOUBLE-HETEROJUNCTION BIPOLAR TRANSISTOR

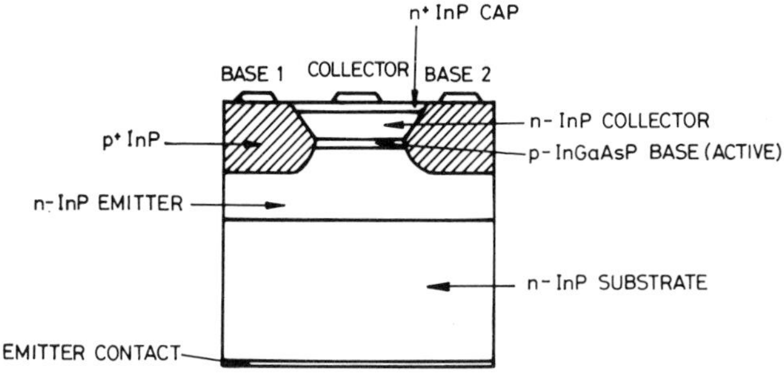

C. DOUBLE HETEROJUNCTION HBT (SHIBATA et. al)

Figure 5.24 Bipolar transistor structures involving (*a*) GaAs–(GaAl)As heterojunction (after Su *et al* [362]), (*b*) InGaAs(P)–InP double heterojunction (after Grote *et al* [365]) and (*c*) InGaAsP–InP double heterojunction as part of an optoelectronic integrated circuit (after Shibata *et al* [332]).

Apart from their compatibility with 1.3–1.6 μm fibre-optic communications systems, these materials offer improved materials properties compared to GaAs such as tunable bandgap, increased high-field drift velocity and higher electron mobilities and diffusion lengths.

Grote *et al* [365] have reported a double-heterojunction bipolar transistor where a wide gap (InP) collector is used instead of the usual homojunction collector (GaInAs) giving the advantages of higher breakdown voltage and reduced leakage currents. The structure is shown in figure 5.24(*b*). The device results showed current gains h_{fe} of up to 320, emitter–collector breakdown voltages (V_{CE_0}) of over 40 V and a cut-off frequency of ~ 1.5 GHz. Shibata *et al* [332] have produced double-heterojunction HBTS using a n-InP (emitter)/p-InGaAsP (base)/n-InP (collector) structure, as part of a small OEIC (figure 5.24(*c*)). The characteristics obtained were $h_{fe} = 400$, $V_{CE_0} = 20$ V, and operating frequency of up to 1.6 GHz. Although these results fall short of expectations for these materials, they probably reflect the less advanced materials technology compared to the GaAlAs/GaAs system.

5.4 Summary

It can be seen from this chapter that liquid-phase epitaxial material has found a wide range of device applications since the technique of LPE was developed in the early 1960s. The earliest devices for which LPE material was used were GaAs Gunn devices and GaP or GaAsP light emitting diodes. The device structures were comparatively simple but exploited the advantages of epitaxy: (i) the ability to control material thickness and carrier concentration, (ii) the possibility of growing p–n junctions and other multilayer structures and (iii) superior electrical, particularly minority carrier, properties due to the lower defect densities in epitaxial material. In this phase, both LPE and conventional VPE were used. In the late 1960s and early 1970s, there was an increasing interest in the III–V ternary alloys, particularly $Ga_{1-x}Al_xAs$ in conjunction with GaAs due to the good lattice matching and the ability of tailoring the bandgap in the alloy and hence producing high-quality heterojunctions. This opened up a new range of devices which exploited the GaAlAs/GaAs interface such as single- and double-heterostructure lasers, photocathodes and solar cells. The fact that Al-containing alloys could not be easily grown by conventional VPE gave the LPE technique a distinct edge in this phase of development. Other alloy systems being studied at this time were $Ga_xIn_{1-x}P$ for blue/green LEDs because of its large direct bandgap of ~ 2.26 eV ($x = 0.74$, 300 K), $Ga_xIn_{1-x}Sb$ for improved Gunn devices due to the postulated increase in the peak-to-valley ratio in the velocity/field characteristics [367], and $Ga_xIn_{1-x}As$ or $GaAs_xSb_{1-x}$ to extend GaAs photocathode response further into the infrared because of the reduced bandgaps of these alloys [368]. At this time the epitaxial growth of the Sb-containing compounds was difficult by vapour-phase techniques due to the instability of SbH_3 as a source of antimony. For the other alloys both LPE and VPE were used depending on the device requirement. The vapour-phase technique was beginning to demonstrate its superiority in terms of precise control of growth rate and good surface topography and hence began to dominate the microwave device area particularly for the more complex structures now being produced. Its ability to perform compositional grading was also being exploited to reduce strain in lattice-mismatched layers [369].

In the mid-1970s, the interest in quaternary III–V alloys began to flourish, particularly in the GaInAsP material lattice matched to InP. These materials gave independent control of lattice parameter and bandgap (and other materials properties). This opened up a new range of heterostructure optoelectronic devices. The excellent compositional control of LPE has made it the favoured technique for growing these quaternary alloys with very low lattice mismatch ($\Delta a/a \sim 10^{-4}$). More recent developments in optoelectronics such as OEICs and buried heterostructure lasers have enabled the LPE technique to exploit its potential for selective area growth and anisotropy of growth rate.

The advances made in the technique of metal-organic vapour-phase epitaxy (MOVPE) since the pioneering work of Manasevit [370] and Bass [371] have seen it become an extremely versatile technique for growing nearly all the III–V materials and a wide range of device structures (optoelectronic, microwave, quantum-well etc). The principal advantages of the technique are (i) the ability to grow complex structures with good control of layer thickness and composition, (ii) the potential for automation and scaling up to production, (iii) the good surface morphology obtained which makes the material suitable for standard photo-lithographic processes for device fabrication and (iv) the ability to grow a wide range of materials over a wide range of growth conditions in a single growth apparatus. This technique is steadily beginning to dominate the field of III–V epitaxy particularly in the research area. There are still however areas where liquid-phase epitaxy retains an edge, particularly for semiconductor lasers and solar cells where luminescence efficiency or minority carrier lifetimes are key parameters, or where the particular abilities of LPE, e.g. for selective area growth, can be exploited.

Chapter 6

Future Developments

6.1 Introduction

The previous chapters have given a history of the LPE technique and its applications to a wide range of electronic devices. The limitations of current LPE technology have been mentioned in these chapters. In the following sections, we shall discuss some of the areas in which future developments in the LPE technique would be beneficial. Also, future trends in III–V materials technology will be briefly discussed together with the likely role of LPE in these areas.

6.2 Future Developments in LPE Technology

The success of LPE growth depends on a large number of factors. Apart from the dominant ones of temperature control, solution composition and substrate preparation, there are a number of subsidiary factors which affect the layer quality and yield, such as the ambient gas purity, the design of the growth equipment (loading systems, furnaces), the method of separation of the solution from the growth layer and the geometry of the growth solution. In the following sections, developments in these areas of LPE technology which could lead to enhancements in material quality are discussed.

6.2.1 Automatic control
There is no doubt that developments in computer technology coupled with the availability of sophisticated, high-quality interface systems have already improved certain aspects of LPE technology, particularly the control and reproducibility of temperature, gas flows and mechanical movements, and the

ability to log large amounts of experimental data for future analysis or to enable repeat runs to be made. The continuing improvements in microprocessor speeds, memory storage capacity and improved interface components (12–16 bit resolution in A/D and D/A converters, low-noise amplifiers, stepper motor drives, etc) should enable further progress to be made in the near future. Of particular benefit has been the use of closed-loop control systems to maintain close control of temperature throughout the lpe growth cycle, by means of the computer taking regular thermocouple readings through A/D converters, comparing the values with the predetermined ones and if necessary sending correction signals through D/A converters to the temperature controllers. This fine control of temperature is of particular importance in the growth of thin layer structures where precise supersaturations must be used. A major limitation in this closed-loop control is the time constant of the furnace and boat system which is a measure of how long it takes for a correction to the temperature control setting to be seen as a change in the boat or crucible temperature. To obtain the maximum benefit from automatic closed-loop control systems, low thermal mass furnaces and boats are required. The use of the gold-plated reflector type of furnace with its low thermal mass is desirable in this type of application. The reduction in thermal mass of boats and crucibles will be discussed in the next section.

6.2.2 Boat and crucible design

Graphite is the most commonly used material for the fabrication of lpe boats and crucibles and is likely to remain so. There has been a continuing improvement in several of the key properties of graphite of importance in this application. Several manufacturers now produce high-purity grades with residual metal contents of ~ 1 ppma. Increases in hardness and flexural strength have enabled more complex boat designs to be made, thinner walls and frames to be used, and higher tolerances on surface flatness and sliding clearances to be specified. Reduced porosity of graphites has reduced the problem of outgassing during lpe growth cycles. Various methods of post-manufacture surface treatment are now available to seal surface pores and produce a high-purity surface coating. These improvements in graphite properties have enabled sliding boats to be designed and manufactured with (i) better wipe-off characteristics, (ii) less wear and hence longer lifetimes, (iii) less outgassing of CO, O_2 and H_2O during the growth cycle, (iv) higher purity, hence reduced backgrounds in the grown layers, (v) reduced thermal mass, which is of importance in automatic growth systems as mentioned in the previous section, and (vi) more complexity. On the latter point, the ability to produce more intricate boat assemblies could lead to several beneficial future developments—the ability to have fine adjustments in the boat to allow for variable shaped substrates or to have multiple substrates in the boat. Both of these developments would enhance the output and reduce the cost of LPE material.

6.2.3 Substrate preparation

As in most types of epitaxy, the surface topography of LPE-grown layers depends critically on the substrate surface—its crystallographic orientation, the magnitude and direction of any misorientation from the low-index plane, the flatness and the presence of surface contamination or lattice damage. The characteristic surface terraces on LPE layers are related to the misorientation and so future improvements in crystal orientation and flatness of polishing should have beneficial effects. The Laué x-ray diffraction technique is the most widely used for single crystal substrate orientation, but its resolution is normally limited to ± 15 arc min. Recent work has led to an improved Laué x-ray orientation technique with a resolution of ± 6 arc s [372], while an optical technique, based on the reflection of laser light from the cleavage planes perpendicular to the substrate surface, can be used to achieve an accuracy of orientation of better than 1 arc min. Substrate wafers orientated to this accuracy are now becoming available from commercial suppliers.

The polishing of semiconductor wafers as substrates for epitaxy is normally a compromise between producing a flat surface and ensuring a minimal depth of damage. Purely mechanical polishing using diamond or alumina preparations used on a wax lap can produce optically flat surfaces on semiconductors but will normally leave a damaged surface region of high dislocation density and possibly embedded particles. Purely chemical

polishing techniques, while giving a defect-free surface, generally produce severe edge rounding, undulating surfaces and preferential etching at defects. Consequently, a variety of 'chemi-
mechanical' techniques have been developed for the III–V
compounds which employ a fine abrasive coupled with an
oxidant such as sodium hypochlorite or hydrogen peroxide.
Even after many years of work by the vendors of polished
semiconductor wafers, there are still problems [373]. This is
partly due to the higher standards of substrate perfection being
demanded by the users of substrates for epitaxy and ion
implantation. Recently, several new approaches to low-damage
semiconductor polishing have been researched, including hydro-
plane [374] or contactless polishing and diamond milling [375].
Further progress on these techniques will be needed in conjunction with the accurate orientation methods mentioned earlier in
order to produce the flatter defect-free substrates that would
significantly improve the quality of LPE layers. This is particularly true of the less common III–V substrate materials such as
GaSb, InAs and InSb which may become increasingly important
in future as substrates for the LPE growth of longer wavelength
laser and detector materials.

6.2.4 Lower-temperature LPE

There are several potential advantages to be gained from
growing LPE films at lower temperatures. The equilibrium concentrations of native defects are lower at reduced growth
temperatures, while the pick up of impurities by the solution
from the LPE boat should be reduced. The solubility of III–V
solutes are reduced as the temperature falls, leading to lower
growth rates which can be of advantage in the growth of thin
layer structures such as quantum-well devices. Finally, the strain
introduced during cool-down from the growth temperature to
room temperature, due to differences in expansion coefficient
between layer and substrate, can be reduced when epitaxy
occurs nearer to room temperature. Other benefits would be
reduced problems with volatile dopants or matrix species and
less interdiffusion at interfaces.

The major limitation to lowering LPE growth temperatures has
been the reduced effectiveness of the removal of impurities in
the growth solution either by evaporation or reaction with the
H_2 ambient gas. The sources of the impurities are (i) solution

starting materials, (ii) boat or crucible, (iii) ambient gas and (iv) leaks in the growth system. The purity of the starting materials, in particular the solvent materials Ga and In which are the majority constituents in LPE growth solutions for III–V epitaxy, has improved steadily over the years to the point where 99.99999% (7 9s) material is available. The solute materials, being synthesised in a high-temperature process, are generally less pure, but form only a small fraction of the solution, typically less than 5 mole per cent. The boat or crucible materials can be a source of impurities both in the matrix and due to adsorbed species. The higher-purity, low-porosity graphites now available have led to improvements, while the use of surface coatings can also be beneficial in reducing the adsorption problem. The use of dry glove boxes for the loading and unloading of LPE boats can also help to reduce problems with adsorbed O_2, H_2O and CO_2 from the atmosphere. Relatively little work has been done on improvements in ambient gas purity. The use of palladium diffused hydrogen is still the standard, although other techniques such as Ga–In–Al eutectic alloys have been used to remove traces of O_2 from the ambient gas [189], and getters have been added to the growth solution [190]. A major problem with many early LPE systems was that of leaks due to the presence of glass-to-glass or glass-to-metal seals using PTFE or Viton materials. More recently, the use of welded joints and UHV-type gasket seals has become more common for the gas handling part of the growth system, although the main seals on the work tube used for loading and unloading are often still a weak point. The use of inert gas blanketing of seals, double O-ring seals, or UHV-type gasket seals would all improve this situation, improving LPE material quality and allowing the use of lower growth temperatures. The benefits of LPE growth in high vacuum have been demonstrated by Stareev [376], with growth temperatures in the range 530–650 °C being used instead of the usual 750 to 800 °C, but this early work has not been followed up, probably due to the increased complexity of the equipment.

6.2.5 Supersaturation

The supersaturation in III–V LPE solutions is normally small since at supercoolings of more than ~ 8 °C, spontaneous homogeneous nucleation is likely to occur. When growing heterostructures, particularly where a large lattice mismatch occurs, an increased supersaturation can have several benefits. Firstly, the increased density of nuclei leads to improved surface morphology due to the earlier coalescence of the nuclei. Also, layers can be grown with larger mismatch to the substrate when the supersaturation is increased. Finally, in cases where the substrate or a previously grown layer is unstable in the growth solution, a high supersaturation can overcome this tendency and allow the new layer to quickly nucleate and cover the underlying material.

There are some papers published on ways to achieve rapid supersaturation of LPE solutions, either by having a step in the furnace temperature profile such that the solution cools as it passes onto the substrate [15] or by cooling the substrate by means of a flow of gas below it [11,18]. However, these early attempts have not been followed up, probably due to difficulties in controlling and reproducing the supercooling. With the recent developments in temperature and flow measurement and control these approaches merit re-investigation. Another approach to rapid changes in supercooling is that of liquid-phase electroepitaxy (LPEE) which was discussed in §2.3.7. Here the cooling at the liquid–solid interface is achieved by the Peltier effect due to current flow through the substrate-solution cell, although the dominant effect is believed to be electromigration. The usefulness of this technique for mismatched systems has been demonstrated by Abul-Fadl *et al* [377] in the case of $In_xGa_{1-x}As$ on GaAs ($x = 0.038$ to 0.054). Further work on this aspect of LPE would be very valuable as it could increase the versatility of the LPE technique.

6.3 Future Developments in III–V Materials Technology

6.3.1 Materials

The immediate future in III–V materials looks certain to be dominated by two materials systems, the GaAs/GaAlAs and InGaAs(P)/InP lattice-matched combinations. Apart from these, there is likely to be increasing interest in lower-bandgap materials involving Sb such as InAsSb, GaInAsSb and InPAsSb driven by (i) the interest in sources and detectors for 2–3 μm wavelength operation in future fibre-optic communication systems and (ii) possible uses in thermal imaging applications in the 3–5 and 8–12 μm wavebands. The main substrates for these alloys are GaSb, InAs and InSb. There are several problems with these materials systems. In the ternary system InAsSb, the large change in lattice constant between InAs and InSb leads to large lattice mismatches when mid-composition range alloys are grown. The large differences in expansion coefficient between InAs and GaSb leads to strain and hence to dislocations when certain layer/substrate combinations are cooled from the growth temperature to room temperature. Added to this is the less-developed technology for substrate polishing and preparation, solution baking schedules, wash melts, doping etc in these low-bandgap materials compared to the wide-bandgap III–Vs. Device technology for these materials is also in a primitive state at present, with little published work on diffusion, ion implantation, contacts and electrical assessment. Much research is thus required on all aspects of the materials and device technology for low-bandgap III–V alloys.

The other possible future development in III–V materials is in the area of quinternary alloys which allow not only the independent choice of bandgap and lattice constant available in the quaternary alloys, but also of expansion coefficient. This parameter is of great importance, since if the expansion coefficient of the layer can be matched to that of the substrate, the introduction of strain during the cool-down from the growth temperature to room temperature can be eliminated. The additional degree of freedom could alternatively be used to select some other composition-dependent property such as refractive index. Mukai *et al* [378] have used LPE to grow $(Al_xGa_{1-x})_{1-z}In_zP_yAs_{1-y}$ lattice matched to GaAs. Excellent agreement between calculated and experimental values of bandgap and lattice constant was found and the layers were uniform in composition with high PL efficiencies. Much research needs to be done on the growth of these five-component alloys to ascertain whether they have useful properties for device applications.

6.3.2 Layer structures

There are three main areas where developments are likely to

occur in III–V device structures in the near future:

(i) growth of device structures on silicon substrates
(ii) selective area growth and growth on profiled substrates
(iii) quantum-well and strained-layer superlattice structures.

In the first category, the interest is in the use of Si substrates because of their availability with large diameters (150 mm) at low cost, compared to GaAs or InP substrates which are expensive and available only with small diameters (50 mm). There is also the possibility of using the Si substrate to carry functional circuitry such as CCDs, FET amplifiers, etc. Most research in this area has been on the growth of GaAs on Si, to combine the optoelectronic and high-frequency microwave properties of GaAs with the large-area availability and advanced signal processing capabilities of silicon. The major problems in obtaining high-quality epitaxial growth of III–Vs on Si are the *in situ* removal of the native oxide from the Si substrate and the mismatch of lattice parameters and expansion coefficients. These problems make it very unlikely that LPE will find any role in this area.

The interest in selective area growth has been driven by the field of integrated optics whereby semiconductor lasers and detectors are integrated on a chip with electronic functions such as FET amplifiers, multiplexers and power control circuitry. The different device structures are grown side by side on various parts of the substrate and the interconnections are made using standard photolithographic techniques or by using the substrate as a common contact. To achieve this type of structure by LPE requires the use of surface masking with SiO_2 or Si_3N_4 with windows opened to define the areas of growth. There will be much interest in future in the use of *in situ* techniques for growing such structures, thus avoiding many stages of photolithography. The main candidate appears to be photolytic MOVPE where the vapour-phase deposition process is controlled by UV illumination of the substrate surface. By the use of non-contact projection imaging techniques, patterns of growth can be defined on the substrate surface. As far as LPE is concerned, there is little possibility of such maskless selective area growth, although recent work using electroepitaxy (LPEE) has shown the possibility of using variations in the thickness of the electrode to produce variations in the growth rate on the substrate. Other than this, selective area LPE will continue to rely on the use of SiO$_2$ surface patterning to define the areas of growth. In the case of LPE, unlike MBE or MOVPE, there is no polycrystalline overgrowth on the mask material, which avoids having to remove the polycrystalline growth by selective etching or a lift-off technique.

The growth of superlattices and quantum-well structures has been achieved by LPE as described in the previous chapter. Control of layer thicknesses in such multilayer structures is difficult since there is a need to make the growth times very short in order to maintain a high enough supersaturation (and hence growth rate) to avoid interface instabilities. Shorter growth times could be achieved by using rotating slider systems with multiple solution wells to enable shorter dwell times to be used. However, there are several limitations of the LPE technique for quantum-well and superlattice structures. Firstly, it is difficult to grade the composition and thickness of the layers as has been done in some strained-layer superlattice structures grown by MOVPE. Secondly, it is hard to maintain the close thickness control in multiple-quantum-well structures due to solute depletion effects and the non-linear growth rate obtained in the 'transient' regime immediately after moving the solution onto the substrate. Thus LPE can be used for small numbers of cycles involving repetition of two lattice-matched materials. However, more complex structures with large numbers of layers, variable composition or thickness layers, or strained layer superlattices with large differences in lattice constant between the individual layers, are likely to be impossible to grow by LPE.

6.4 Summary

Although there are several areas where improvements in LPE technology would enable better-quality material and more complex structures to be grown with improved control, it is likely that the liquid-phase technique will have a limited capability in terms of increasingly sophisticated device structures being envisaged for the future. However, there appears to be a secure future for LPE material in certain areas of III–V electronic devices, particularly where the device structure is reasonably simple or where economies dictate a lower-technology growth technique.

Bibliography

J J Hsieh 1980 in *Handbook on Semiconductors* vol. 3 ed. S P Keller (North-Holland) p. 415 (*General review*)

L R Dawson 1972 in *Progress in Solid State Chemistry* vol. 7 ed. H Reiss and J O McCaldin (Pergamon) p. 117 (*GaAs and GaAlAs*)

K Nakajima 1982 in *GaInAsP Alloy Semiconductors* ed. T P Pearsall (Wiley) p. 43 (*GaInAsP*)

K W Benz and E Bauser 1980 in *Crystals—Growth, Properties and Applications* vol 3 ed. H C Freyhardt (Springer) p. 1 (*Growth mechanisms*)

P D Greene 1987 in *Advanced Crystal Growth* ed. P M Dryburgh *et al* (Prentice-Hall) p. 221 (*General review*)

J C Brice 1980 in *Current Topics in Materials Science* vol. 5 ed. E Kaldis (North-Holland) p. 99 (*Theory, techniques, GaAs, GaAlAs*)

M B Panish and M Ilegems 1972 in *Progress in Solid State Chemistry* vol. 7 ed. H Reiss and J O McCaldin (Pergamon) p. 39 (*III–V phase diagrams*)

Special edition of *Journal of Crystal Growth* on liquid-phase epitaxy, vol. 27, December 1974

Proceedings of the International Symposia on GaAs and Related Compounds (Institute of Physics Conference Series vols 3, 7, 9, 17, 24, 33a, 33b, 45, 56, 65, 74, 83, 91 and 96)

References

[1] Royer L 1928 *Bull. Soc. Fr. Miner.* **51** 7

[2] Nelson H 1963 *RCA Rev.* **24** 603

[3] Hall R N 1963 *J. Electrochem. Soc.* **110** 385

[4] Hurle D T J 1979 *J. Phys. Chem. Solids* **40** 613

[5] Venables J A and Price G L 1975 *Epitaxial Growth* part B ed. J W Mathews (Academic)

[6] Stowell M J 1974 *J. Crystal Growth* **24/25** 45

[7] van der Merwe J H 1970 *J. Appl. Phys.* **41** 4725

[8] Faust J W and John H F 1962 *J. Phys. Chem. Solids* **23** 1119

[9] Longo J T, Gertner E R and Harris J S 1973 *J. Nonmetals* **1** 321

[10] Sangster R 1962 *Compound Semiconductors* vol 1 ed. R K Willardson and H L Goering (New York: Reinhold)

[11] Longo J T, Harris J S, Gertner E R and Chu J C 1972 *J. Crystal Growth* **15** 107

[12] Astles M G and Rowland M C 1974 *J. Crystal Growth* **27** 142

[13] Crossley I and Small M B 1973 *J. Crystal Growth* **19** 160

[14] Blom G M 1971 *J. Electrochem. Soc.* **18** 1834

[15] Moon R L and van der Plas H A 1978 *J. Electron. Mater.* **7** 347

[16] Ohta K, Kawashima M, Tanuma T and Kataoka S 1977 *Japan. J. Appl. Phys.* **16** 1605

[17] Astles M G, Dosser O D, Maclean A J and Wright P J 1981 *J. Crystal Growth* **54** 485

[18] Mattes B L and Route R K 1972 *J. Crystal Growth* **16** 219

[19] Saul R H and Roccasecca D D 1973 *J. Appl. Phys.* **44** 1983

[20] Bauser E, Frik M, Loechner K S, Schmidt L and Ulrich R 1974 *J. Crystal Growth* **27** 148

[21] Mottram A and Peaker A R 1974 *J. Crystal Growth* **27** 193

[22] Peters R C 1972 *4th Int. Symp. on GaAs and Related Compounds (Boulder, Colorado)* (Inst. Phys. Conf. Ser. No 17) p 55

[23] Pak K, Nishinaga T and Uchiyama S 1977 *Japan. J. Appl. Phys* **16** 949

[24] Rode D L 1975 *Phys. Status Solidi* a **32** 425

[25] Holmes D E and Kamath G S 1980 *J. Electron. Mater.* **9** 95
[26] Woelk C and Benz K W 1974 *J. Crystal Growth* **27** 177
[27] Wrick V, Scilla G J, Eastman L F, Henry R L and Swiggard E M 1976 *Electron. Lett.* **12** 394
[28] Kawakami T and Sugiyama K 1973 *Japan. J. Appl. Phys.* **12** 1808
[29] Ishihara O, Otsubo M and Mitsui S 1977 *Japan. J. Appl. Phys.* **16** 2109
[30] Takahashi S and Nagai H 1981 *J. Crystal Growth* **51** 502
[31] Ettenberg M and McFarlane III S H 1974 *J. Crystal Growth* **23** 233
[32] Bauser E and Strunk H P 1984 *J. Crystal Growth* **69** 561
[33] Keller K and Muench W V 1971 *Solid State Electron.* **14** 526
[34] Rowland M C and Maclean A J 1976 unpublished work at RSRE Baldock
[35] Rosztoczy F E, Ermanis F, Hayashi I and Schwartz B 1970 *J. Appl. Phys.* **41** 264
[36] Constantinescu C and Petrescu-Prahova I 1967 *J. Phys. Chem. Solids* **28** 2397
[37] Vilms J and Garrett J P 1972 *Solid State Electron.* **15** 443
[38] Harris J S and Snyder W L 1969 *Solid State Electron.* **12** 337
[39] Solomon R 1968 *2nd Int. Symp. on GaAs and Related Compounds (Dallas, Texas)* (Inst. Phys. Conf. Ser. No 7) p 11
[40] Kang C S and Greene P E 1968 *2nd Int. Symp. on GaAs and Related Compounds (Dallas, Texas)* (Inst. Phys. Conf. Ser. No 7) p 18
[41] Greene P D 1971 *Solid State Commun.* **9** 1299
[42] Goodwin A R, Dobson C D and Franks J 1968 *2nd Int. Symp. on GaAs and Related Compounds (Dallas, Texas)* (Inst. Phys. Conf. Ser. No 7) p 36
[43] Casey Jr H C, Panish M B and Wolfstirn K B 1971 *J. Phys. Chem. Solids* **32** 571
[44] Astles M G 1975 unpublished results
[45] Woodward D W, Kirchner P D, Schaff W J, Tiwari S, Stall R and Eastman L F 1980 *8th Int. Symp. on GaAs and Related Compounds (Vienna)* (Inst. Phys. Conf. Ser. No 56) p 83
[46] Astles M G, Smith F G H and Williams E W 1973 *J. Electrochem. Soc.* **120** 1750
[47] Abrams E B, Sumski S, Bonner W A and Coleman J J 1979 *J. Appl. Phys.* **50** 4469
[48] Wada O, Mayerfeld A and Robson P N 1980 *J. Electrochem. Soc.* **127** 2278
[49] Umeba I and Robson P N 1981 *J. Crystal Growth* **53** 292
[50] Baumann G G, Benz K W and Pilkuhn M H 1976 *J. Electrochem. Soc.* **123** 1232
[51] Rosztoczy F E, Antypas G A and Casau C J 1970 *3rd Int. Symp.*

on GaAs and Related Compounds (Aachen) (Inst. Phys. Conf. Ser. No 9) p 86

[52] Brown K E 1974 *Solid State Electron.* **17** 505

[53] Jordan A S, Trumbore F A, Wolfstirn K B, Kowalchik M and Rocasecca D D 1973 *J. Electrochem. Soc.* **120** 791

[54] Saul R H and Hackett W H 1970 *J. Appl. Phys.* **41** 3554

[55] Sudlow P D, Mottram A and Peaker A R 1972 *J. Mater. Sci.* **7** 168

[56] Pearsall T P 1980 *IEEE J. Quantum Electron.* **QE-16** 709

[57] Feng M, Tashima M M, Cook L W and Stillman G E 1978 *7th Int. Symp. on GaAs and Related Compounds (St Louis)* (Inst. Phys. Conf. Ser. No 45) p 61

[58] Small M B, Blackwell J C and Potemski R M 1979 *J. Crystal Growth* **46** 253

[59] Rosztoczy F E, Goldwasser R E and Kinoshita J 1974 *5th Int. Symp. on GaAs and Related Compounds (Deauville)* (Inst. Phys. Conf. Ser. No 24) p 37

[60] Kim C K, Malbon R M, Omori M and Park Y S 1978 *7th Int. Symp. on GaAs and Related Compounds (St Louis)* (Inst. Phys. Conf. Ser. No 45) p 305

[61] Sasai Y, Hase N and Kajiwara T 1985 *Japan. J. Appl. Phys.* **24** L137

[62] Astles M G 1974 *PhD Thesis* University of Oxford

[63] Panish M B, Hayashi I and Sumski S 1969 *IEEE J. Quantum Electron.* **5** 210

[64] Ruppecht H 1966 *1st Int. Symp. on GaAs and Related Compounds (Reading)* (Inst. Phys. Conf. Ser. No 3) p 57

[65] Astles M G, Birbeck J C H, Laversuch C J and Rowland M C 1976 *J. Crystal Growth* **34** 24

[66] Panish M B, Hayashi I and Sumski S 1970 *Appl. Phys. Lett.* **16** 326

[67] Minden H T 1974 *J. Crystal Growth* **23** 79

[68] Aylett M R, Faktor M M, Haigh J and White E A D 1981 *J. Crystal Growth* **54** 604

[69] Otsubo M, Segawa K and Miki H 1973 *Japan. J. Appl. Phys.* **12** 797

[70] Harding W R 1970 unpublished work at RSRE Baldock

[71] Thompson G H B and Kirkby P A 1974 *J. Crystal Growth* **27** 70

[72] Horikoski Y 1976 *Japan. J. Appl. Phys.* **15** 887

[73] Dawson L R 1977 *J. Appl. Phys.* **48** 2485

[74] Heinen J 1982 *J. Crystal Growth* **58** 596

[75] Dutt B V, Roccasecca D D, Temkin H and Bonner W A 1984 *J. Crystal Growth* **66** 525

[76] Reed T B 1969 *Solid State Research Report (1969:1)*, MIT Lincoln Laboratory, MA, USA

[77] Morkoc H and Eastman L F 1976 *J. Crystal Growth* **36** 109

[78] Fukui T and Kobayashi T 1977 *Japan. J. Appl. Phys.* **16** 2081

[79] Ishii M, Kau H and Susaki W 1976 *Appl. Phys. Lett.* **29** 375

[80] Nishizawa J, Okuno Y and Yoshida S 1975 *IEEE Trans. Electron Devices* **ED-22** 716

[81] Nishizawa J, Okuno Y and Tadano H 1975 *J. Crystal Growth* **31** 215

[82] Hsieh J J 1974 *J. Crystal Growth* **27** 49

[83] Nelson R J 1979 *Appl. Phys. Lett.* **35** 654

[84] Yamamoto T, Sakai K and Akiba S 1977 *Japan. J. Appl. Phys.* **16** 1699

[85] Pollack M A, Nahory R E, DeWinter J C and Ballman A A 1978 *Appl. Phys. Lett.* **33** 314

[86] Feng M, Cook L W, Tashima M M, Windhorn T H and Stillman G E 1979 *Appl. Phys. Lett.* **34** 292

[87] Wright P D, Rezek E A and Holonyak Jr N 1977 *J. Crystal Growth* **41** 254

[88] Oe K and Sugiyama K 1978 *Appl. Phys. Lett.* **33** 449

[89] Bolkhovityanov Y B and Zembatov H B 1977 *J. Crystal Growth* **37** 101

[90] Kano H, Miyazawa S and Sugiyama K 1979 *Japan. J. Appl. Phys.* **18** 2183

[91] Sankaran R, Antypas G A, Moon R L, Escher J S and James L W 1976 *J. Vac. Sci. Technol.* **13** 932

[92] Pearsall T P and Hopson Jr R W 1977 *J. Appl. Phys.* **48** 4407

[93] Sakai K, Matsushima Y, Akiba S and Yamamoto T 1979 *Japan. J. Appl. Phys.* **18** 1009

[94] Nahory R E, Pollock M A, DeWinter J C and Williams K M 1977 *J. Appl. Phys.* **48** 1607

[95] Hersee S D, Carter A C, Goodfellow R C, Hawkins G and Griffith I 1979 *Solid State Electron Devices* **3** 179

[96] Hsieh J J, Finn M C and Rossi J A 1976 *6th Int. Symp. on GaAs and Related Compounds (St Louis)* (Inst. Phys. Conf. Ser. No 33a) p 37

[97] Nakajima K, Kusunoki T, Komiya S, Akita K and Kotani T 1977 *24th Spring Meeting of Japan. Soc. of Appl. Phys. (Tokyo)*

[98] Kano H and Sugiyama K 1980 *Electron. Lett.* **16** 146

[99] Kobayashi N, Horikoshi Y and Uemara C 1979 *Japan. J. Appl. Phys.* **18** 2169

[100] Long S I, Ballantyne J M and Eastman L F 1974 *J. Crystal Growth* **26** 13

[101] Davies R 1976 unpublished work (Plessey Research Ltd UK)

[102] Tiller W A 1968 *J. Crystal Growth* **2** 69

[103] Deitch R 1970 *J. Crystal Growth* **7** 69

[104] Lawrence D J and Eastman L F 1975 *J. Crystal Growth* **30** 267

[105] Kumagawa M, Witt A F, Lichtensteiger M and Gatos H C 1973 *J. Electrochem. Soc.* **120** 583

[106] Daniele J J 1975 *Appl. Phys. Lett.* **27** 373

[107] Imamura Y, Jastrzebski L and Gatos H C 1979 *J. Electrochem. Soc.* **126** 1381

[108] Blom G M, Daniele J J, Kyros T and Witt A F 1975 *J. Electrochem. Soc.* **122** 1541

[109] Imamura Y, Jastrzebski L and Gatos H C 1978 *J. Electrochem. Soc.* **125** 1560

[110] Jastrzebski L, Imamura Y and Gatos H C 1978 *J. Electrochem. Soc.* **125** 1140

[111] Pfann W G, Benson K E and Wernick J H 1957 *J. Electron.* **2** 597

[112] Daniele J J and Hebling A J 1981 *J. Appl. Phys.* **52** 4325

[113] Tiller W A and Kang C 1968 *J. Crystal Growth* **2** 345

[114] Small M B and Barnes J F 1969 *J. Crystal Growth* **5** 9

[115] Minden H T 1970 *J. Crystal Growth* **6** 228

[116] Rode D L 1973 *J. Crystal Growth* **20** 13

[117] Crossley I and Small M B 1972 *J. Crystal Growth* **15** 268

[118] Moon R L 1974 *J. Crystal Growth* **27** 62

[119] Ghez R 1973 *J. Crystal Growth* **19** 153

[120] Cook L W 1981 *J. Electron. Mater.* **10** 119

[121] Crossley I and Small M B 1971 *J. Crystal Growth* **11** 157

[122] Pak K, Nishinaga T and Nakamura T 1979 *Japan. J. Appl. Phys.* **18** 1699

[123] Ahn B H, Shurtz R R and Trussel C W 1971 *J. Appl. Phys.* **42** 4512

[124] Pearsall T P, Quillec M and Pollock M A 1979 *Appl. Phys. Lett.* **35** 342

[125] Nakajima K, Tanahashi T, Akita K and Yamaoka T 1979 *J. Appl. Phys.* **50** 4975

[126] Bolkhovityanov Yu B and Zembatov H B 1977 *J. Crystal Growth* **37** 101

[127] Chand N, Syrbu A V and Houston P A 1983 *J. Crystal Growth* **61** 53

[128] Panish M B and Ilegems M 1972 *Progress in Solid State Chemistry* vol 7 (New York: Pergamon)

[129] Panish M B and Arthur J R 1970 *J. Chem. Thermodynamics* **2** 299

[130] Astles M G 1974 *J. Chem. Thermodynamics* **6** 105

[131] Nakajima K, Kusunoki T and Akita K 1980 *Fujitsu Sci. Tech. J.* **16** 59

[132] Kuphal E 1984 *J. Crystal Growth* **67** 441

[133] Dedegkaev T T, Kryukov I I, Lideikis T P, Tsarenkov B V and Yakovlev Yu P 1978 *Sov. Phys.–Tech. Phys.* **23** 350

[134] Ilegems M and Pearson G L 1968 *Proc. 2nd Int. Symp. on GaAs and Related Compounds (Dallas, Texas)* (Inst. Phys. Conf. Ser. No 7) p 9
[135] Nakajima K, Osamura K, Yasuda K and Nurakami Y 1977 *J. Crystal Growth* **41** 87
[136] Panish M B and Sumski S 1969 *J. Phys. Chem. Solids* **30** 129
[137] Hsieh J J 1981 *IEEE J. Quantum Electron.* **QE-17** 118
[138] Nagai H and Noguchi Y 1978 *Appl. Phys. Lett.* **32** 234
[139] Tabatabaie-Alavi K, Perea E H and Fonstad C G 1981 *J. Electron. Mater.* **10** 591
[140] Vieland L J 1963 *Acta Metall.* **11** 137
[141] Stringfellow G B 1974 *J. Crystal Growth* **27** 21
[142] Antypas G A and Edgecumbe J 1976 *J. Crystal Growth* **34** 132
[143] Perea E H and Fonstad C G 1980 *J. Electrochem. Soc.* **127** 313
[144] Lazarev V B and Dashevskii 1965 *Inorg. Mater.* **1** 1901
[145] Dashevskii M, Kukuladze G V, Lazarev V B and Mirgalovskaya M S 1967 *Inorg. Mater.* **3** 1561
[146] Glazov V M, Chizhevskaya S N and Glagoleva N N 1969 *Liquid Semiconductors* (New York: Plenum)
[147] Szapiro S 1980 *J. Phys. Chem. Solids* **41** 279
[148] Jordan A S 1970 *Met. Trans.* **1** 239
[149] Linnebach R 1979 *ECCG-2 Conf. (Lancaster, UK) September 1979*
[150] Bolkhovityanov Yu B and Chikichev S I 1983 *Crystal Res. Technol.* **18** 847
[151] Konig U 1977 *J. Electrochem. Soc.* **124** 1414
[152] Schauer K-H and Vogel A 1970 *3rd Int. Symp. on GaAs and Related Compounds (Aachen)* (Inst. Phys. Conf. Ser. No 9) p 100
[153] James L W, Antypas G A, Uebbing J J, Yep T O and Bell R L 1971 *J. Appl. Phys.* **42** 580
[154] Moon R L 1980 *J. Appl. Phys.* **51** 5561
[155] Andrews A M, Cheung D T, Gertner E R and Longo J T 1976 *J. Vac. Sci. Technol.* **13** 961
[156] Bubulac L O, Andrews A M, Gertner E R and Cheung D T 1980 *Appl. Phys. Lett.* **36** 734
[157] Ilegems M and Panish M B 1973 *J. Crystal Growth* **20** 77
[158] Onton A, Lorenz M R and Reuter W 1971 *J. Appl. Phys.* **42** 3420
[159] Campbell J C, Hitchens W R, Holonyak Jr N, Lee M H, Ludowise M J and Coleman J J 1974 *Appl. Phys. Lett.* **24** 327
[160] Rupprecht H, Woodall J M and Pettit G D 1967 *Appl. Phys. Lett.* **11** 81

[161] Nakajima K, Tanahashi T and Akita K 1982 *Appl. Phys. Lett.* **41** 194

[162] Wu T Y and Pearson G L 1972 *J. Phys. Chem. Solids* **33** 409

[163] Pearsall T P, Bisaro R, Ansel R and Merenda P 1978 *Appl. Phys. Lett.* **32** 497

[164] Joullie A, Dedies R, Chevrier J and Bougnot G 1974 *Rev. Physique Appl.* **9** 455

[165] Miki H, Segawa K, Otsubo M, Shirahata K and Fujibayashi K 1978 *Japan. J. Appl. Phys.* **17** 2079

[166] Ilegems M and Panish M B 1974 *J. Phys. Chem. Solids* **35** 409

[167] Brown R L and Sobers R G 1974 *J. Appl. Phys.* **45** 4735

[168] Kagawa T and Motosugi G 1979 *Japan. J. Appl. Phys.* **18** 1001

[169] Antypas G A and Moon R L 1974 *J. Electrochem. Soc.* **121** 416

[170] Nakajima K and Akita K 1981 *J. Crystal Growth* **54** 232

[171] Swarup P, Jain R K, Verma S N, Charan S and Tandle D M 1982 *Phys. Status Solidi* a **72** K189

[172] Kazamura M, Ohta I and Teramoto I 1983 *Japan. J. Appl. Phys.* **22** 654

[173] Antypas G A, Moon R L, James L W, Edgecumbe J and Bell R L 1972 *4th Int. Symp. on GaAs and Related Compounds (Boulder, Colorado)* (Inst. Phys. Conf. Ser. No 17) p 48

[174] Mukai S, Matsuzaki M and Shimada J 1980 *Japan. J. Appl. Phys.* **19** L505

[175] Fujimoto A, Watanabe H, Takeuchi M and Shimura M 1985 *Japan. J. Appl. Phys.* **24** L653

[176] Shimura M, Fujimoto A, Yasuda H and Yamashita S 1982 *Japan. J. Appl. Phys.* **21** L338

[177] Sankaran R and Antypas G A 1976 *J. Crystal Growth* **36** 198

[178] DeWinter J C, Pollock M A, Srivastava A K and Zyskind J L 1985 *J. Electron. Mater.* **14** 729

[179] Gertner E R, Cheung D T, Andrews A M and Longo J T 1977 *J. Electron. Mater.* **6** 163

[180] Nagai H and Noguchi Y 1976 *J. Appl. Phys.* **47** 5484

[181] Lendvay E 1982 *Electron. Lett.* **18** 407

[182] Skromme B J, Low T S and Stillman G E 1982 *10th Int. Symp. on GaAs and Related Compounds (Albuquerque)* (Inst. Phys. Conf. Ser. No 65) p 485

[183] Nanishi Y 1978 *Japan. J. Appl. Phys.* **17** 1177

[184] Houng Y M, Pearson G L and Mattes B L 1978 *J. Electrochem. Soc.* **125** 2058

[185] Morkoc H, Eastman L F and Woodward D 1980 *Thin Solid Films* **71** 245

[186] Chang S C, Meng G Y and Stevenson D A 1983 *J. Crystal*

Growth **62** 465
[187] Takahashi K 1980 *Japan. J. Appl. Phys.* **19** 775
[188] Mattes B L, Houng Y M and Pearson G L 1975 *J. Vac. Sci. Technol.* **12** 869
—→ [189] Shealy J R and Woodall J M 1982 *Appl. Phys. Lett.* **41** 88
[190] Stevenson D A, Ketrush P I, Chang S C and Borshchevsky A 1980 *Appl. Phys. Lett.* **37** 832
—→ [191] Hicks H G B and Greene P D 1970 *3rd Int. Symp. on GaAs and Related Compounds (Aachen)* (Inst. Phys. Conf. Ser. No 9) p 92
[192] Kuphal E and Pocker A 1982 *J. Crystal Growth* **58** 133
[193] Ip K T, Eastman L F and Wrick V L 1977 *Electron. Lett.* **13** 682
[194] Groves S H and Plonko M C 1978 *7th Int. Symp. on GaAs and Related Compounds (St Louis)* (Inst. Phys. Conf. Ser. No 45) p 71
[195] Cook L W, Tabatabaie N, Tashima M M, Windhorn T W, Bulman G E and Stillman G E 1980 *8th Int. Symp. on GaAs and Related Compounds (Vienna)* (Inst. Phys. Conf. Ser. No 56) p 361
[196] Holmes D E and Kamath G S 1981 *J. Crystal Growth* **54** 51
[197] Bhattacharya P K, Rao M V and Tsai M-J 1983 *J. Appl. Phys.* **54** 5096
[198] Beneking H, Grote N and Selders J 1981 *J. Crystal Growth* **54** 59
[199] Otsubo M and Miki H 1975 *Japan. J. Appl. Phys.* **14** 621
[200] Otsubo M, Segawa K and Miki H 1973 *Japan. J. Appl. Phys.* **12** 797
[201] Wolfe C M, Stillman G E and Lindley W T 1970 *J. Appl. Phys.* **41** 3088
[202] Ashley K L, Carr D L and Romano-Moran R 1973 *Appl. Phys. Lett.* **22** 23
[203] Allenson M and Bass S J 1976 *Appl. Phys. Lett.* **28** 3
[204] Wight D R, Oliver P E, Prentice T and Steward V W 1981 *J. Crystal Growth* **55** 183
[205] Nelson R J 1978 *7th Int. Symp. on GaAs and Related Compounds (St Louis)* (Inst. Phys. Conf. Ser. No 45) p 256
[206] Ryan R D and Eberhardt J E 1972 *Solid State Electron.* **15** 865
[207] Sekela A M, Feucht D L and Milnes A G 1974 *5th Int. Symp. on GaAs and Related Compounds (Deauville)* (Inst. Phys. Conf. Ser. No 24) p 245
[208] Shirafuji J, Tamura A, Inoue M and Inuishi Y 1981 *J. Appl. Phys.* **52** 4704
[209] Booker G R, Titchmarsh J M, Fletcher J, Darby D B, Hockley M and Al-Jassim M 1978 *J. Crystal Growth* **45** 407
[210] Lang D V and Logan R A 1975 *J. Electron. Mater.* **4** 1053

[211] Mitonneau A 1977 *Electron. Lett.* **13** 666
[212] Vasudev P K, Mattes B L and Bube R H 1978 *7th Int. Symp. on GaAs and Related Compounds (St Louis)* (Inst. Phys. Conf. Ser. No 45) p 154
[213] Li S S, Lin C Y, Bedair S M and Hutchby J A 1982 *J. Electron. Mater.* **11** 273
[214] Hasegawa F and Mayerfeld A 1975 *Electron. Lett.* **11** 286
[215] Okumura T and Ikoma T 1978 *J. Crystal Growth* **45** 459
[216] Chiao S H and Antypas G A 1978 *J. Appl. Phys.* **49** 466
[217] Williams E W and Blacknall D M 1967 *Trans. Met. Soc. AIME* **239** 387
[218] Ashen D J, Dean P J, Hurle D T J, Mullin J B, White A M and Greene P D 1975 *J. Phys. Chem. Solids* **36** 1041
[219] Fukui T and Horikoshi Y 1979 *Japan. J. Appl. Phys.* **18** 961
[220] Rogulin V Y and Shlenskii A A 1973 *Sov. Phys.–Semicond.* **7** 673
[221] Shih K and Pettit G D 1974 *J. Electron. Mater.* **3** 391
[222] Dapkus P D, Hackett Jr W H, Lorimor O G and Bachrach R Z 1974 *J. Appl. Phys.* **45** 4920
[223] Hamilton B, Peaker A R, Bramwell S, Harding W and Wight D R 1975 *Appl. Phys. Lett.* **26** 12
[224] Rozgonyi G A, Petroff P M and Panish M B 1974 *Appl. Phys. Lett.* **24** 251
[225] Kumar K and Takagi M 1977 *Japan. J. Appl. Phys.* **16** 957
[226] Rowland M C and Smith D A 1977 *J. Crystal Growth* **38** 143
[227] Oe K, Shinoda Y and Sugiyama K 1978 *Appl. Phys. Lett.* **33** 962
[228] Bauser E and Strunk H P 1984 *J. Crystal Growth* **69** 561
[229] Small M B, Blakeslee A E, Shih K K and Potemski R M 1975 *J. Crystal Growth* **30** 257
[230] Astles M G, Hill H and Steward V 1983 *J. Crystal Growth* **62** 61
[231] Novotny J, Srobar F, Moravec F and Zelinka J 1983 *J. Crystal Growth* **65** 596
[232] Nishizawa J and Okuno Y 1978 *Technical Report TR-41* Research Institute of Electrical Communication, Tohoku University, Sendai, Japan
[233] Keramides V G, Temkin H and Bonner W A 1982 *Appl. Phys. Lett.* **40** 731
[234] Besomi P, Degani J, Dutta N K, Wagner W R and Nelson R J 1984 *J. Appl. Phys.* **56** 2879
[235] Sakai K, Akiba S and Yamamoto T 1977 *Japan. J. Appl. Phys.* **16** 2043
[236] Nelson A W, Westbrook L D and White E A D 1982 *J. Crystal Growth* **58** 236
[237] Rezek E A, Holonyak N, Vojak B A, Stillman G E, Rossi J A,

Keune J L and Fairing J D 1977 *Appl. Phys. Lett.* **31** 288

[238] Rezek E A, Vojak B A, Chin R, Holonyak N and Samman E A 1981 *J. Electron. Mater.* **10** 255

[239] Capella R M, Benchimol J L, Bresse J F and Quillec M 1984 *11th Int. Symp. on GaAs and Related Compounds (Biarritz)* (Inst. Phys. Conf. Ser. No 74) p 163

[240] Zwicknagl P, Rehm W and Bauser E 1984 *J. Electron. Mater.* **13** 545

[241] Benchimol J L, Clempkes S, N'Guyen D C, LeRoux G, Bresse J F and Primot J 1986 *J. Appl. Phys.* **59** 4068

[242] Nakajima K, Yamazaki S and Akita K 1982 *Japan. J. Appl. Phys.* **21** L237

[243] Matsumoto Y, Taguchi K and Ishida K 1984 *J. Crystal Growth* **69** 53

[244] Cook L W, Feng M, Tashima M M, Blattner R J and Stillman G E 1980 *Appl. Phys. Lett.* **37** 173

[245] Feng M, Cook L W, Tashima M M and Stillman G E 1979 *Appl. Phys. Lett.* **34** 697

[246] Brunemeier P E, Roth T J, Holonyak Jr N and Stillman G E 1984 *J. Appl. Phys.* **56** 1707

[247] Wakefield B 1978 *Appl. Phys. Lett.* **33** 408

[248] Wilson R B, Besomi P and Nelson R J 1985 *J. Electrochem. Soc.* **132** 172

[249] Dingle R 1977 *J. Vac. Sci. Technol.* **14** 1006

[250] Coleman J J and Nash F R 1978 *Electron. Lett.* **14** 558

[251] Tabatabaie-Alavi K, Markunas R J and Fonstad C J 1982 *J. Electrochem. Soc.* **129** 2085

[252] Takeda Y, Kondo M and Sasaki A 1984 *J. Crystal Growth* **66** 475

[253] Whitney P S and Fonstad C G 1985 *J. Appl. Phys.* **57** 4663

[254] Fujita S, Kuzuhara M, Yagyu M and Sasaki A 1982 *Solid State Electron.* **25** 359

[255] Garner C M, Su C Y and Spicer W E 1979 *J. Vac. Sci. Technol.* **16** 1521

[256] Griffiths R J M, Chew N, Cullis A G and Joyce G C 1983 *Electron. Lett.* **19** 988

[257] Konig U and Yorke H 1985 *J. Crystal Growth* **73** 515

[258] Tsang W T, Logan R A and van der Ziel J P 1982 *Appl. Phys. Lett.* **40** 942

[259] Turley S E H, Henshall G D, Greene P D, Knight V P, Moule D J and Wheeler S A 1981 *Electron. Lett.* **17** 868

[260] Turley S E H and Greene P D 1982 *J. Crystal Growth* **58** 409

[261] Kirkby P A and Thompson G H B 1976 *J. Appl. Phys.* **47** 4578

[262] Feldman R D and Austin R F 1985 *J. Crystal Growth* **71** 1

[263] DeCremoux B 1978 *7th Int. Symp. on GaAs and Related Compounds (St Louis)* (Inst. Phys. Conf. Ser. No 45) p 52

[264] Varon J, Mahieu M, Vanderberg P, Boissy M-C and Lebailly J 1981 *IEEE Trans. Electron Devices* **ED-28** 416

[265] Nishizawa J, Koibe M and Jin C C 1983 *J. Appl. Phys.* **54** 2807

[266] Ishiguro H, Sawa K, Nagao S, Yamanaks H and Koike S 1983 *Appl. Phys. Lett.* **43** 1034

[267] Tsai M-J and Wu C H 1984 *11th Int. Symp. on GaAs and Related Compounds (Biarritz)* (Inst. Phys. Conf. Ser. No 74) p 439

[268] Carter A C, Goodfellow R C and Davis R 1980 *Int. Conf. on Communications (Seattle) June 1980* (New York: IEEE)

[269] Hall R N, Fenner G E, Kingsley J D, Soltys T J and Carlson R O 1962 *Phys. Rev. Lett.* **9** 366

[270] Hayashi I, Panish M B and Reinhart F K 1971 *J. Appl. Phys.* **42** 1929

[271] Hsieh J J, Rossi J A and Donnelly J P 1976 *Appl. Phys. Lett.* **28** 709

[272] Logan R A, Temkin H, Merritt F R and Mahajan S 1984 *Appl. Phys. Lett.* **45** 1275

[273] Petroff P M and Hartman R L 1973 *Appl. Phys. Lett.* **23** 469

[274] Hutchinson P W and Dobson P S 1975 *Philos. Mag.* **32** 745

[275] Mahajan S, Johnson Jr W D, Pollock M A and Nahory R E 1979 *Appl. Phys. Lett.* **34** 717

[276] Mahajan S 1983 *Proc. 3rd Microscopy of Semiconducting Materials Conf. (Oxford)* p 259

[277] Jacob G, Duseaux M, Farges J P, Van Den Boom M M B and Roksnoer P J 1983 *J. Crystal Growth* **61** 417

[278] Seki Y, Matsui J and Watanabe H 1976 *J. Appl. Phys.* **47** 3374

[279] van der Ziel J P, Logan R A, Nordland W A and Kazarinov R F 1985 *J. Appl. Phys.* **57** 1759

[280] Yonezu H, Sakuma I, Kobayashi K, Kamejima T, Ueno M and Nannichi Y 1973 *Japan. J. Appl. Phys.* **12** 1585

[281] Tsang W T, Logan R A and Johnson L F 1979 *J. Appl. Phys.* **50** 5165

[282] Doi A, Fukuzawa T, Nakamura M, Ito R and Aiki K 1979 *Appl. Phys. Lett.* **35** 441

[283] Nelson A W, Westbrook L D and Evans J S 1983 *Electron. Lett.* **19** 34

[284] Burkland H, Kuphal E and Dinges H W 1986 *Electron. Lett.* **22** 802

[285] Temkin H, Dolan G J, Logan R A, Kazarinov R F, Olsson N A and Henry C H 1985 *Appl. Phys. Lett.* **46** 105

[286] Schilling M and Wünstel K 1986 *Appl. Phys. Lett.* **49** 710

[287] Liau Z L and Walpole J N 1982 *Appl. Phys. Lett.* **40** 568
[288] Kinoshita J, Okuda H and Uematsu Y 1983 *Electron. Lett.* **19** 215
[289] Nahory R E, Pollack M A, Beebe E D, DeWinter J C and Dixon R W 1976 *Appl. Phys. Lett.* **28** 19
[290] Dolginov L M, Drakin A E, Druzhinina L V, Eliseev P G, Mil'vidskii M G, Sverdlov B M and Skripkin V A 1983 *Trudy FIAN* **141** 349 (Engl. transl. 1984 *J. Sov. Laser Res.* **5** 349)
[291] Caneau C, Srivastova A K, Dentai A G, Zyskind J L and Pollack M A 1985 *Electron. Lett.* **21** 815
[292] Dingle R, Gossard A C and Wiegmann W 1975 *Phys. Rev. Lett.* **34** 1327
[293] Esaki L and Chang L L 1974 *Phys. Rev. Lett.* **33** 495
[294] Kolbas R M, Holonyak Jr N, Dupuis R D and Dapkus P D 1978 *Sov. Tech. Phys. Lett.* **4** 28
[295] Dupuis R D and Dapkus P D 1977 *Appl. Phys. Lett.* **31** 366
[296] Holonyak Jr N, Kolbas R M, Laidig W, Vojak B A, Dupuis R D and Dapkus P D 1978 *7th Int. Symp. on GaAs and Related Compounds (St Louis)* (Inst. Phys. Conf. Ser. No 45) p 387
[297] Rezek E A, Vojak B A and Holonyak Jr N 1978 *J. Appl. Phys.* **49** 5398
[298] Rezek E A, Chin R, Holonyak Jr N, Kirchoefer S W and Kolbas R M 1979 *Appl. Phys. Lett.* **35** 45
[299] Dutta N K, Napholtz S G, Yen R, Brown R L, Shen T M, Olsson N A and Craft D C 1985 *Appl. Phys. Lett.* **46** 19
[300] Sasai Y, Hase N, Ogura M and Kajiwara T 1985 *12th Int. Symp. on GaAs and Related Compounds (Karuizawa)* (Inst. Phys. Conf. Ser. No 79) p 709
[301] Leheny R F, Nahory R E and Pollock M A 1979 *Electron. Lett.* **15** 713
[302] Lee T P, Burrus C A, Dentai A G and Ogawa K 1980 *Electron. Lett.* **16** 156
[303] Armiento C A, Donnelly J P and Groves S H 1979 *Appl. Phys. Lett.* **34** 229
[304] Bowers J E, Srivastava A K, Burrus C A, DeWinter J C, Pollock M A and Zyskind J L 1986 *Electron. Lett.* **22** 137
[305] Mohammed K, Capasso F, Logan R A, van der Ziel J P and Hutchinson A L 1986 *Electron. Lett.* **22** 215
[306] Tabatabaie N, Lepkowski T R, Cook L W and Stillman G E 1982 *10th Int. Symp. on GaAs and Related Compounds (Albuquerque)* (Inst. Phys. Conf. Ser. No 65) p 217
[307] Wieder H H, Clawson A R and McWilliams G E 1977 *Appl. Phys. Lett.* **31** 468
[308] Washington M A, Nahory R E, Pollock M A and Beebe E D

1978 *Appl. Phys. Lett.* **33** 854

[309] Schemmel G, Dorn P, Hess K, Linnebach R and Lösch K 1982 *10th Int. Symp. on GaAs and Related Compounds (Albuquerque)* (Inst. Phys. Conf. Ser. No 65) p 209

[310] Gertner E R, Andrews A M, Bubulac L O, Cheung D T, Ludowise M J and Riedel R A 1979 *J. Electron. Mater.* **8** 545

[311] Sukegawa T, Hiraguchi T, Janaka A and Hagino M 1978 *Appl. Phys. Lett.* **32** 376

[312] Capasso F, Panish M B, Sumski S and Foy P W 1980 *Appl. Phys. Lett.* **36** 165

[313] Chin R and Hill C M 1982 *Appl. Phys. Lett.* **40** 332

[314] Susa N, Nakagome H, Mikami O, Ando H and Kanbe H 1980 *IEEE J. Quantum Electron* **QE-16** 864

[315] Trommer R and Hoffman L 1984 *11th Int. Symp. on GaAs and Related Compounds (Biarritz)* (Inst. Phys. Conf. Ser. No 74) p 409

[316] Capasso F and Tsang W T 1984 *AT&T Bell Laboratories Record* (March 1984)

[317] Tomasetta L R, Law H D, Eden R C, Deyhimy I and Nakano K 1978 *IEEE J. Quantum Electron.* **QE-14** 800

[318] Gammel J C and Ballantyne J M 1980 *Appl. Phys. Lett.* **36** 149

[319] Gammel J C, Metze G and Ballantyne J M 1979 *Proc. IEDM* (New York: IEEE) p 635

[320] Gammel J C, Ohno H and Ballantyne J M 1981 *IEEE J. Quantum Electron.* **QE-17** 269

[321] Chen C Y, Cox H M, Garbinski P A and Hummel S G 1984 *Appl. Phys. Lett.* **45** 867

[322] Degani J, Leheny R F, Nahory R E, Pollock M A, Heritage J P and DeWinter J C 1981 *Appl. Phys. Lett.* **38** 27

[323] Chen C Y and Chi G C 1984 *Appl. Phys. Lett.* **45** 1083

[324] Cazarre A, Tasselli J, Marty A, Bailbe J P and Rey G 1985 *Electron. Lett.* **21** 1124

[325] Takeda Y, Hamaguchi N, Kondo M, Fujita S and Sasaki A 1984 *3rd Record III–V Alloy Semiconductor Physics and Electronics Seminar* p 157

[326] Chand N, Houston P A and Robson P N 1985 *Electron. Lett.* **21** 308

[327] Ury I, Margalit S, Yust M and Yariv A 1979 *Appl. Phys. Lett.* **34** 430

[328] Wada O, Hamaguchi H, Miura S, Makiuchi M, Yamakoshi S, Sakurai T, Nakai K and Iguchi K 1983 *Electron. Lett.* **19** 1031

[329] Matsueda H, Saski S and Nakamura M 1983 *J. Lightwave Technol.* **LT-1** 368

[330] Leheny R F, Nahory R E, Pollack M A, Ballman A A, Beebe E

D, DeWinter J C and Martin R J 1980 *Electron. Lett.* **16** 353

[331] Koren U, Yu K L, Chen T R, Bar-Chaim N, Margalit S and Yariv A 1982 *Appl. Phys. Lett.* **40** 643

[332] Shibata J, Nakao I, Sasai Y, Kimura S, Hase N and Serizawa H 1984 *Appl. Phys. Lett.* **45** 191

[333] Mori Y, Shibata J, Sasai Y, Serizawa H and Kajiwara T 1985 *12th Int. Symp. on GaAs and Related Compounds (Karuizawa)* (Inst. Phys. Conf. Ser. No 79) p 661

[334] Scheer J J and VanLaar J 1965 *Solid State Commun.* **3** 189

[335] Turnbull A A and Evans G B 1968 *J. Phys. D: Appl. Phys.* **1** 155

[336] Allenson M B, King P G R, Rowland M C, Steward G J and Syms C H A 1972 *J. Phys. D: Appl. Phys.* **5** L89

[337] Antypas G A and Edgecumbe J 1975 *Appl. Phys. Lett.* **26** 371

[338] Allenson M B and Bass S J 1976 *Appl. Phys. Lett.* **28** 113

[339] Woodall J M and Hovel H J 1977 *J. Crystal Growth* **39** 108

[340] Sahai R, Edwall D D and Harris J S 1979 *Appl. Phys. Lett.* **34** 147

[341] Gavand M, Mayet L and Laugier A 1984 *11th Int. Symp. on GaAs and Related Compounds (Biarritz)* (Inst. Phys. Conf. Ser. No 74) p 469

[342] Moon R L, James L W, VanderPlas H A and Nelson N J 1978 *Appl. Phys. Lett.* **33** 196

[343] Rosztoczy F E and Kinoshita J 1974 *J. Electrochem. Soc.* **121** 439

[344] Rosztoczy F E, Goldwasser R E and Kinoshita J 1974 *5th Int. Symp. on GaAs and Related Compounds (Deauville)* (Inst. Phys. Conf. Ser. No 24) p 37

[345] Nanishi Y, Takahei K and Kuroiwa K 1978 *J. Crystal Growth* **45** 272

[346] Tiwari S, Spencer M G, Levy H M, Judaprawira S, Woodward D W and Eastman L F 1980 *8th Int. Symp. on GaAs and Related Compounds (Vienna)* (Inst. Phys. Conf. Ser. No 56) p 475

[347] Li C, Chen P T and Wang P H 1978 *7th Int. Symp. on GaAs and Related Compounds (St Louis)* (Inst. Phys. Conf. Ser. No 45) p 353

[348] Schmitt R, Steiner K, Kaufmann L M F, Brockerhoff W and Heime K 1985 *12th Int. Symp. on GaAs and Related Compounds (Karuizawa)* (Inst. Phys. Conf. Ser. No 79) p 619

[349] Morkoc H, Andrews J T, Houng Y M, Sankaran R, Bandy S G and Antypas G A 1978 *Electron. Lett.* **14** 448

[350] Kasahara K, Hayashi J and Nomura H 1984 *Electron. Lett.* **20** 618

[351] Rosztoczy F E, Long S I and Kinoshita J 1974 *J. Crystal Growth* **27** 205

[352] Kim C K, Malbon R M, Omori M and Park Y S 1978 *7th Int. Symp. on GaAs and Related Compounds (St Louis)* (Inst. Phys. Conf. Ser. No 45) p 305

[353] Knight S, Dawson L R, Di Lorenzo J V and Johnson W A 1970 *3rd Int. Symp. on GaAs and Related Compounds (Aachen)* (Inst. Phys. Conf. Ser. No 9) p 108

[354] Rode D L, Schwartz B and Di Lorenzo J V 1974 *Solid State Electron.* **17** 1119

[355] Wrick III V L and Eastman L F 1974 *5th Int. Symp. on GaAs and Related Compounds (Deauville)* (Inst. Phys. Conf. Ser. No 24) p 31

[356] Astles M G and Williams E W 1972 *Electron. Lett.* **8** 2120

[357] Fairman R D, Omori M and Fank F B 1976 *6th Int. Symp. on GaAs and Related Compounds (St Louis)* (Inst. Phys. Conf. Ser. No 33a) p 45

[358] Yen K H and Berenz J J 1982 *Electron. Lett.* **18** 171

[359] Roszoczy F E, Caldwell J F, Kinoshita J and Omori M 1973 *Appl. Phys. Lett.* **22** 525

[360] Beneking H, Su L and Ponse F 1982 *Microelectron. J.* **13** 5

[361] Konagai M, Katsukawa K and Takahashi K 1977 *J. Appl. Phys.* **48** 4389

[362] Su L M 1982 *10th Int. Symp. on GaAs and Related Compounds (Albuquerque)* (Inst. Phys. Conf. Ser. No 65) p 423

[363] Rey G, Bailbe J P, Marty A, Antarasana C, Cazarre A and Tasseli J 1985 *11th Int. Symp. on GaAs and Related Compounds (Biarritz)* (Inst. Phys. Conf. Ser. No 74) p 671

[364] Su L M, Grote N, Bach H G, Doldissen W and Rosenweig M 1984 *11th Int. Symp. on GaAs and Related Compounds (Biarritz)* (Inst. Phys. Conf. Ser. No 74) p 605

[365] Grote N, Su L M and Bach H-G 1985 *12th Int. Symp. on GaAs and Related Compounds (Karuizawa)* (Inst. Phys. Conf. Ser. No 75) p 583

[366] Zipperian T E, Dawson L R and Barnes C E 1982 *10th Int. Symp. on GaAs and Related Compounds (Albuquerque)* (Inst. Phys. Conf. Ser. No 65) p 523

[367] Hilsum C and Rees H D 1970 *Electron. Lett.* **6** 277

[368] James L W, Antypas G A, Uebbing J J, Edgecumbe J and Bell R L 1970 *3rd Int. Symp. on GaAs and Related Compounds (Aachen)* (Inst. Phys. Conf. Ser. No 9) p 195

[369] Abrahams M S, Weisberg L R, Buiocchi C J and Blanc J 1969 *J. Mater. Sci.* **4** 223

[370] Manasevit H M and Simpson W I 1969 *J. Electrochem. Soc.* **116** 1725

[371] Bass S J 1975 *J. Crystal Growth* **31** 172

210   *LPE Growth of III–V Compounds for Devices*

[372] Gillies D C 1987 *J. Electron. Mater.* **16** 151
[373] Silva R M, Orazio F D and Bennett J M 1986 *Optics News* **12** 10
[374] Gormley J V, Manfra M J and Calawa A R 1981 *Rev. Sci. Instrum.* **52** 1256
[375] Mroczkowski J A and Vydyanath H R 1981 *J. Electrochem. Soc.* **128** 655
[376] Stareev G D 1976 *J. Crystal Growth* **32** 189
[377] Abul-Fadl A, Stefanakos E K and Collis W J 1981 *J. Crystal Growth* **51** 279
[378] Mukai S, Yajima H, Mitsuhashi Y, Yanagisawa S and Kutsuwada N 1984 *Appl. Phys. Lett.* **44** 904
[379] Rubenstein M 1962 *Electrochem. Soc. Abs.* **11** 129
[380] Ilegems M, Panish M B and Arthur J R 1974 *J. Chem. Thermodynamics* **6** 157
[381] Onton A and Chicotka R J 1970 *J. Appl. Phys.* **41** 4205

Index

For information on the various III–V ternary and quaternary alloys, look in the index under the name of the group III element with the highest atomic number in the alloy.